Letters from Africa

◆

Travel Stories from an Adventuresome Soul

Cynthia Wales Tuthill

iUniverse, Inc.
New York Bloomington

Letters from Africa

Travel Stories from an Adventuresome Soul

Copyright © 2008 by Cynthia W. Tuthill

All rights reserved. No part of this book may be used or reproduced by any means, graphic, electronic, or mechanical, including photocopying, recording, taping or by any information storage retrieval system without the written permission of the publisher except in the case of brief quotations embodied in critical articles and reviews.

iUniverse books may be ordered through booksellers or by contacting:

iUniverse
1663 Liberty Drive
Bloomington, IN 47403
www.iuniverse.com
1-800-Authors (1-800-288-4677)

Because of the dynamic nature of the Internet, any Web addresses or links contained in this book may have changed since publication and may no longer be valid. The views expressed in this work are solely those of the author and do not necessarily reflect the views of the publisher, and the publisher hereby disclaims any responsibility for them.

ISBN: 978-0-595-53123-3 (pbk)
ISBN: 978-0-595-63184-1 (ebk)

Printed in the United States of America

iUniverse rev. date: 10/29/08

To Kalpana

You have made my life a more beautiful place.
Our relationship continues, even after you left this world
to become a shining meteor.

"*The wheat, which is golden, will remind me of you.
And I'll love the sound of the wind in the wheat …*"

Antoine de Saint-Exupery
"The Little Prince."

Contents

To Kalpana v
Acknowledgments: ix
Introduction xi
Chapter 1:   Africa 2003 1
Chapter 2:   Africa 2004 21
Chapter 3:   Africa 2005 55
Chapter 4:   Africa 2006 91
Chapter 5:   Africa 2007 155

Acknowledgments:

I owe a debt of gratitude to Michelle Slade, for the countless hours she spent providing her exquisite editorial skills to make my letters more manageable for readers.

My deepest thanks to Sunita and Deepa: my Didis.

To Ellen and Ron, my appreciation for your wonderful friendsip and being brave enough to join us on a trip so far from home.

I give a toast of honor to my beloved mother and father, for raising me to have open eyes and heart, and providing unconditional love.

To my daughter Jessica and my son Mark, I can never express the depth of joy and pride I have experienced in being your mother, and for the abiding friendship and enduring love that I feel for you. And I give my deep appreciation also to Adam and Stephanie, for loving my children and becoming a treasured part of our family.

And most importantly to Jim, my soulmate, the light of my life … without whom none of this would have been possible … truly, you are necessary to me.

Introduction

In February 2003 my dearest friend Kalpana Chawla was killed in the tragic crash of the space shuttle Columbia. During the various memorial services, I began to get to know her sisters. I was acquainted with them, of course, as members of the family of my best friend, but they lived far away in India. Whenever they had visited Kalpana here in the U.S., I tended to make myself scarce so that she could spend more time with her visitors.

Yet, this bond that connected us now, of deep unremitting grief, was strong. I loved to gaze at Kalpana's elder sisters Sunita and Deepa as they reminded me of her. And they loved to be with me, her friend, for the same reason. Tentatively, we got to know one another, talking about our lives, our families, our love of birding and nature, and our favorite subject: dear Kalpana.

My husband Jim and I had made plans to spend our summer vacation that year with Kalpana. As I was devastated by having to cancel those plans, he suggested that we go somewhere really special – anywhere at all. We chose Africa, a mysterious and wondrous place we'd read so much about and had always wanted to visit.

And so, the start of another lifelong relationship …

When we returned from this amazing trip of a lifetime, I called Sunita to let her know how magnificent it had been. She was enchanted and breathed, "Tell me ALL about it!" … which was impossible to respond to in a short discussion. This was before the invention of Skype, so that phone call would have been quite expensive too.

So instead of talking for hours, I decided to write her an email about the trip. I titled it "Africa 2003," and sent it off.

We so deeply loved our time in Africa that we have returned every year. And each year I wrote about the trips, with Sunita as my muse. Through these stories, we have developed a treasured friendship.

I am sharing those stories here.

Chapter 1
Africa 2003

◆

Africa is now embedded in our hearts!

Both Jim and I felt the same way ... we deeply loved the continent. During our entire time there, we felt as if we were standing where life itself began, with the wide open skies making us feel as if the whole vista of the universe was ours to explore.

Incredible wildlife. Inspiring hope for the future. Exciting times ... relaxing times ... romantic times. Quite a trip!

I should start at the beginning. Both Jim and I had each always wanted to go to Africa. But we feared that it was too dangerous now with all the human unrest, and perhaps too sad with the extensive animal poaching. Last year however, we met some really nice women from South Africa who told us that the safari business was still going strong, that the safari camps are really nice ... and that there are still animals in Africa! That allowed us to begin dreaming of a trip.

On February 1, when we heard that the shuttle had crashed, and I realized I'd lost a soul mate, all I could think was "I can't go on."

It took months before I was able to talk about where we should go for our summer vacation. I had canceled the plans that we'd made to travel with Kalpana, and it was difficult to consider something new.

One day in late March, however, Jim correctly judged that I was ready for the conversation. He suggested our dream place, Africa. The idea sparkled. Later that same day my dad called me, out of the blue, to tell me he'd gone to a slide show the night before given by some friends who run a travel agency. Specializing in planning flying trips throughout Africa.

Fate works in funny ways…

I zipped over to dad's house to get the brochure and called the company, Bushtracks, the next day. They planned the whole trip for us and it was absolutely perfect. Flawless. First, we had to choose between different parts of Africa. I originally thought of east Africa - Kenya, Mt. Kilimanjaro, Tanzania - but those areas are becoming quite crowded with tourists and now they are in political turmoil. Then there is also South Africa including Cape Town, which is supposed to be quite nice, but we wanted to see animals. So southern Africa was recommended - Botswana and Zambia. Zimbabwe is also not a good place to travel now as it is also in political difficulties.

I also want to make a comment about the cost of the trip. Bushtracks plans safaris which are rather expensive, because for that extra money you can be assured of one-on-one interaction with guides, smaller groups, more remote camps, and therefore a more intimate interaction with the animals. We are also more certain that our money is contributing directly to communities in remote African areas. This doesn't mean that less expensive safaris aren't also wonderful, just that we made this choice based on our limited time and our desire to give as much back to the local communities as possible.

With the itinerary set, Bushtracks sent us a complete planning kit including all the info we needed about inoculations (yes we got quite a few … ouch … plus malaria pills of a new kind which don't lead to insanity like the older ones sometimes did), a little pouch to keep our itinerary and vouchers in, and even some really nice books on Africa. Jim spent weeks reading about the proper clothing and gear and bought us all the khaki and tan clothing we could ever need. They eventually knew us at the local outfitting store, the Redwood Trading Post, "Oh here's the dynamic duo again. What is it you need this time? Safari hats? Gloves? Zip-off pants?" All the planning paid off as we had exactly the right clothes and blended in with the scenery.

And speaking of gear, Jim was in charge of all the camera stuff. He brought along a small pocket camera (a Canon PowerShot), which turned out to be quite convenient for our walking trips and quick shots where we didn't want to pull out a large camera. In fact, a great majority of our best photos were taken with this camera. He also loved using it with our spotting scope (Swarovski), to do "digiscoping" (taking a photo through the scope). This combo gave us our best close-up bird shots. Jim likes fancy equipment, so he did bring along a larger Canon digital SLR camera, with various zoom and telephoto lenses (his favorite was the 70 – 200 mm). We also purchased image-stabilized binoculars, which were extremely useful, along with straps (I preferred a "harness," where the straps go over the shoulders so the binocular weight isn't focused on the back of the neck). We loved having a digital video camera to capture the sounds and motion of Africa. We found one with low-light capability, which was great for night shots. And we made sure not to forget the rechargeable batteries, chargers, and extra video tapes! We later put the best photos on a website, pbase.com/tuthillx2, to share with everyone.

August arrived and it was time to go. The actual trip had several different stages. First there was the travel stage. This was hard. An 11 hour flight to London, an 8 hour layover, and then another 11 hour flight to Johannesburg (referred to almost exclusively as "Joburg"). It was quite grueling! We did manage to have a nice vegetarian Indian lunch at our "favorite" restaurant in London. Since we've only been to London together once for three days I guess it's rather pretentious to claim that we HAVE a "favorite" restaurant in London but it felt fun for us to return to a place we really did love when we were in London that previous time. The restaurant closes at 3pm and we landed at 1:40pm, so the timing was tight. We had to rush to get through customs, catch the Heathrow express, take the underground to the Bond station and then hustle down Oxford Street against the teeming masses of shoppers. Dressed in our safari clothes, sweating profusely (a record heat wave in London that day) and pushing our way through the crowds, we probably presented quite a picture. We arrived at the restaurant at 2:55pm and they were gracious enough to give us the full treatment, staying open late.

Back at Heathrow, we began a new custom: playing travel Scrabble while enjoying vodka and chocolate, waiting for our next flight.

Our travel agent had brilliantly planned a 24-hour "rest stop" in Joburg before starting our first safari. We were met after our 9am arrival at the airport baggage claim area by a rather oldish white-haired but affable man holding a sign with "Tuthill X 2." Jim looked at me with raised eyebrows - what happened to the "Orr" – his last name – in this family?! We were transported to our hotel, the Grace Rosebank. During the 45 minute ride and again the next morning returning to the airport our driver regaled us with tales of South Africa, pointing out all the important landmarks with nonstop commentary. The hotel was quite comfortable ... even fancy ... with a huge suite of rooms including a marble bathroom. The hot bath we immediately took was marvelous. We spent most of the day shopping in the outdoor market that had stall after stall of local handicrafts. We enjoyed eating there too, contrary to our travel book's insistence that one should never eat food from "local street vendors." Oops. But it was delicious! We had our first cup of "bush tea" (Rooibos or red bush) which became a daily ritual. We bought trinkets and a lovely piece of painted cloth to use as a futon cover in our guest room back home.

At long last, we were headed for "Xakanaxa" (pronounced "kakanaka"), our first safari camp in Botswana, in the Okavango Delta. I'd recently read out loud to Jim a series of fiction books set in Botswana about a woman who owns a detective agency in Gabarone and since we simply adored the books (very sweet and romantic) we felt a real excitement about being in Botswana. In fact we wanted to MEET Ma Ramotswe, who alas is only a fictional character. So we felt like kids on Christmas Eve as we flew in a small plane to the tiny town of Maun and then in a really tiny 6-seater plane to a dirt strip in the Moremi Game Preserve within the Delta. As we flew over the strip before landing Jim spied the first animals of the trip: elephants! And they were ON the end of the runway. Great excitement! I didn't see them until we had landed, signed in at the ranger station, and hopped into the open-topped Land Rover that was there to meet us. We found our names on the roster: "Tuthill X 2," but by now Jim was becoming used to being just a number. The guide who met us introduced himself as "Leahbo" as we zoomed down the runway and pulled over in front of the elephants. Behind us the plane took off and I realized that the truck was positioned such in order to prevent the elephants from wandering onto the runway. Quite thrilling!

A half hour drive through winding bumpy dirt roads took us to camp. The foliage was quite varied, from low bushes to towering trees surrounded by tall golden grasses. The dry season in southern Africa lasts from May to November with literally never a single drop of rain. So by August things are becoming dry yet many of the deciduous trees have not lost their leaves so there is still quite a bit of greenery. I was surprised to see so many tall trees. For some reason I only ever pictured the wide-open grasslands of the Serengeti when I visualized Africa so I was pleasantly surprised by the intermingling of grasslands and woodlands and thickets. And the many grey solid towering termite mounds which would become a common sight.

Leahbo stopped the jeep to point out a group of impala. We were thrilled to see our first antelopes! We took pictures, and video, and "ooohed" and "aaahed." We appreciated that Leahbo talked about them as if they were also awesome to him. By the end of three days we'd seen so many impala that we eventually didn't even glance their way and so it was considerate that he would realize that since we had just arrived, an impala would be interesting to us at that moment. As we grew in knowledge he stopped pointing out the things that we were already accustomed to. From that perspective it worked well that this camp "assigns" a single guide to each family as he can watch your progress and remember which animals you have never seen and which you may be becoming cavalier about!

We saw more elephants and stopped to watch them browsing the trees. I was amazed at how close we could come to these gigantic gentle animals without their being noticeably uncomfortable. In fact, they just turned and looked our way briefly and then went back to what they were doing. After reading "The Eye of the Elephant," by Delia and Mark Owens, which described the disastrous situation with poaching of elephants in the national parks in Zambia from 1986 to 1991, I was ecstatic to see these elephants being content to be near to people. This means that the poaching has decreased to such an extent that elephants no longer live in fear. Wonderful!

We then arrived at our bush camp on a small river of the Okavango delta system. Like all the camps we stayed in, it was centered on a main room or "sitenge" where the meals were served and people could "hang out" and tell stories. In this camp the main area included not only a large dining room with a great long table but also an expansive deck stretching along the bank of the river, even projecting out over the river, dotted with comfortable chairs. There was also a bar area, library, and a small pool for dipping. This deck area was open but covered with the delicate branches from several tall trees including a "sausage tree" that has dangling fruit shaped like a sausage, about 4 inches in diameter and 12 inches long … wild looking! All of these trees were filled with birds and I was immediately finding new birds I'd never seen before, jotting them down in our book to keep track. I think we must have found 10 new birds in the first few minutes. Our favorites were the little "black crakes" that lived under part of the decking. They looked like tiny versions of a coot, black and round and puffy but with brilliant red eyes, red feet with immense long toes, and a bright orange beak. These adorable little guys scooted around and ate scraps right out of our hands. We also loved the African jacana bird … blue, brown, and white, and again with very long toes. It's called the "Jesus bird" as it appears to walk on water with those long toes. We enjoyed repeating the names of the other birds on the deck: black-eyed bulbuls, black-collared barbets, and white-rumped babblers.

And there were many baobab trees! Anyone who has read "The Little Prince" would love to see these stately old venerable trees. They look so implausible and yet so magnificent.

A young crocodile often could be found basking in the sun right next to one of the decks. In fact he was usually to be found next to a sign that proclaimed "beware of crocodiles." Apparently there were also some very LARGE crocodiles nearby, 10 to 15 feet long. The deck had steps leading down to a grassy patch on the river bank and we were told it was NOT ok to go off the deck and onto the grass as they apparently move very quickly.

After a profuse welcome by the staff we were shown to our tent. We followed the hostess along a dirt walkway beside the river and under more beautiful trees. And then, heaven on earth! Our tent was situated right on the river, with a wooden floor of soft old boards that extended out in front of the tent as a verandah. Two comfortable lounging chairs sat on the deck and inside, an invitingly soft bed with white fluffy down comforter, bedside tables with candles, a writing desk and a dresser with a pitcher of fresh cool water. Behind the tent was an attached bathroom enclosed with bamboo walls but without a roof.

It was SO wonderful taking outdoor showers and enjoying birding while brushing teeth! Jim saw our first "Go-Away" birds up in the trees above our loo. The grey lourie whines in a way that sounds like "go away …," and is accused of having cried this when men were trying to shoot wild animals. As legend goes, these birds would perch above targeted animals and warn them that trouble was coming.

Our tent had open sides with netting and we were surrounded by trees and foliage offering us complete privacy, yet we could sleep almost out in the open, listening to the very loud sounds of hippos and elephants in the night. The hippos were mostly bellowing while the elephants we could hear shaking the trees to remove the leaves and breaking the branches and munching. We were able to walk around the camp during the day but at night we needed an escort … just in case we came across an elephant or hippo or buffalo and didn't know what to do.

We stayed at this camp for three nights and it was marvelous.

Each morning we got up before sunrise, had tea and biscuits, and then started out for a morning game drive in the Land Rovers. We were paired with a nice German couple who didn't really talk much which was fine as the experience was mostly about seeing the animals. Leahbo told us what we'd want to know about each animal: its habits and ecological niche. As the animals are not frightened by the vehicles, which have become part of their landscape, we could get quite close to them. Often we would be only a few feet away from the particular animal we were watching, even lions.

One day we went for a "game drive" on the river instead, in a small motorboat. It was really beautiful and relaxing meandering between the waving grasses of the delta, enjoying more gorgeous birds, especially the African darters with their long graceful necks that looked like snakes sticking up out of the water. We also took a trip in a "mokorro," the small dug-out canoes that were previously used in the river system of the delta and steered by a man standing in back pushing with a pole. We saw otters swimming in the water and it was incredible to be so low on the water. The solitude was comforting.

On the game drives we enjoyed large groups of zebra (the ones in Botswana have a brown stripe between the black ones) and baboons, vervet monkeys and giraffes, and of course more hippos and elephants, many impala, waterbucks and the water-loving antelopes called lechwe. And lions! We came across a pride of lions who had just gorged on a hippo. One male was still eating his fill. There were several lionesses lying about with full stomachs and we saw one give another a lovely caress with her head. One of the lionesses had several cubs congregated around her … one lying on his back, legs sprawled in a perfect posture of satiation. Another cub was nursing and would periodically look over at us. They were sweet and innocent looking. In fact, most of the animal groups we saw had tiny babies - we watched zebra, giraffe, elephants, baboons and monkeys nursing their young.

Another time we came to a pool of water that had about 40 crocodiles around the edges basking in the sun. It seemed eerie to see so many of them. We saw trees filled with vultures, and baboons picking through each other's coats or running around with little babies hanging on their stomachs.

Each day after the morning game drive, during which we always stopped for a tea/coffee break with more biscuits, we returned to camp for a large scrumptious brunch. This was followed by a "rest time" when we hung out in our tent together napping or reading or talking. I wrote in my journal sitting out on the verandah. The sound of drumming would break our reverie at 3pm, and after more tea and cookies we would take off for our afternoon game drive. But not before the most important question was asked: what would you like for "sundowners"? This is the beverage enjoyed at sunset, which occurred at about 6:00 each evening. At this camp we started with red wine but after a few days we discovered "Amarula," a liqueur made from the fruit of the marula tree and which elephants also love. It tastes a bit like Baileys Irish Cream. We fell in love with the yummy flavor and switched our sundowners to Amarula nearly every day. After the evening hours of game viewing and a quick drive back to camp we'd have another drink followed by an incredibly delicious four-course dinner (including special vegetarian dishes prepared just for me) at the long table in the common room.

It was wonderful to be in the Okavango delta. It's an amazing natural wonder. The river begins in Angola and the river course originally drained down to the ocean. But then there was a shifting of the ground - this is after all part of the great rift system - and the waters were trapped. So each wet season (November to May) the rains fill the river and it flows down to this section of the Kalahari Desert, filling in more and more of the waterways. While we were visiting in August, some parts of the river that Leahbo referred to as "new water" had only just filled in recently. It seemed miraculous to me that new rainwater was still coming down the river even though the rains stopped months ago. By the end of the dry season many of the waterways have dried up to wait, dormant, for the next year's rain. During that dry time a large number of the animals who had ventured into the Kalahari Desert during the wet season would return to the delta to find water.

Jim and I had not only read about the delta itself and the desert but also about the Bushman, the most original tribe in Africa. They have a remarkable culture; I was moved by their spiritual beliefs as well as the cleverness required to live in such an inhospitable terrain as the Kalahari Desert. Everything we experienced in Botswana was highlighted by an appreciation for this incredible indigenous culture.

The guides that we had at all our camps were brilliant and exceptionally well trained. They knew EVERY animal, insect, and plant, each of their interrelationships and habits, and all of which they were able to discuss in English. This was impressive. They knew all the calls of all the animals. We would be zipping along the dirt track in the Land Rover and suddenly come to a screeching halt. The guide had either seen some animals which he or she would then point out to us, or had heard the call of an animal or bird. He would then help locate the animal or bird for us. With this expert help we saw 113 species of birds which we'd never seen before. Imagine!

The guides really taught me to listen more than I'd been accustomed to. Not only just the obvious sounds of Africa, which were loud and cacophonous and constant, and which we fell in love with, but also the particular calls of birds. I was so proud of myself when I suddenly noticed that I had begun to "subtract out" the ubiquitous sounds of the various doves and was able to hear a single other bird calling against that background clamor. The doves, by the way, did have rather repetitious calls. One guide told us that Cape turtle dove was calling, "work harder … work harder … work harder …" in the morning but then "drink longer … drink longer … drink longer …" at sundown! And the red-eye dove was calling, "Red eyes! What a pity. Red eyes! What a pity. Red eyes! What a pity," while the emerald-spotted dove, which is a gorgeous bird, with its pitiful long drawn out call was saying, "my mother is dead my father is dead my family is dead … poor me … poor me … poor me … poor me …"

Both of us felt that the night sounds of Africa … the hippos grunting and the elephants munching and the lions roaring … were part of what made us feel as if we were really, truly and finally at home. I experienced deep pleasure each time I awoke at night listening. As expressed by Laurens Van der Post in "The Heart of the Hunter, "I have always been grateful that I was born into a world … where the lion is still roaring. Heard in his native setting it is for me the most beautiful sound in the world. It is to silence what the shooting star is to the dark of the night."

Before each drive we were filled with excitement and during each drive we were never disappointed. Each trip had some new "find" such as a lone hyena guiltily skulking along or a group of skittish wart hogs leaning down on their front knees to eat and looking as if they are missing a pair of pants.

And of course the birds! Splendid multi-colored lilac breasted rollers falling like leaves from the sky; diving screeching fish eagles, even more majestic and regal than bald eagles; clown-colored ox-peckers on the backs of hippos and giraffes prancing about and pecking for food; white-crowned plovers with their huge yellow wattles following us everywhere and yelling at us to stay away from their nests.

On the subject of animals, there were essentially no BUGS in Botswana or Zambia even though we were staying right next to rivers in every camp. This was probably because it was the dry season, but it was certainly nice. We didn't really need the quarts and quarts of repellent which we'd packed. Since we were south of the equator and it was winter the weather was perfect. Beautiful clear blue skies greeted us each day and the temperature hovered about 75 degrees. It was never hot and actually, a bit cold during some of the evening and morning game drives. At night the stars were brilliant and we thrilled at seeing the Southern Cross. And each night the moon grew fuller and fuller …

After our stay of several days it was time to say goodbye to Botswana. Following a final morning of game viewing we were driven back to the little dirt strip where our plane was waiting. Just like clock work. This 6-seat plane took us to a town on the northern border called Kasane. From there we were met by a guy with a truck and again, the "Tuthill X 2" sign. We were driven to the Zambezi River which forms the border with Zambia, Zimbabwe and Namibia, the "four corners." There, another fellow met us with a small boat and zipped us and our luggage across the river. On the opposite river bank a van was waiting and drove us to our hotel in Livingstone at the top of Victoria Falls. The driver of the van took care of getting our visas for a 10 day stay in Zambia. All went quite smoothly.

They say you have to take the bad with the good, well, about now is when the "bad" part of our trip ensued. I'll try not to harp on it too much. The hotel we stayed in was the Royal Livingstone, and was supposed to represent the glamour of a by-gone day. We thought that meant the glamour of exploration, like David Livingstone, for example. Unfortunately, we now know it meant the "glamour" of colonialism. The staff at the Livingstone acted subservient, which was a very uncomfortable experience for us. We did not like staying in a hotel with other guests who would want to be treated that way. Furthermore, after being out in the bush, living with and listening to the wild animals, it was horrid to be trapped in a hotel room with no open windows. There were no screens and since there were a lot of inquisitive monkeys about, we were told to keep our sliding glass doors shut. We couldn't even hear the waterfall during the night. For us, it was really tragic.

But, back to the good.

The Falls.

Or "Mosi-ao-Tunya" as the locals called it - The Smoke that Thunders.

Wow. A fabulous, awe-inspiring, astounding cascade, raging over steep black rock cliffs, creating enormous clouds of mist and bright vertical rainbows. A life-altering, sublime vision.

It is said that when David Livingstone first saw the Falls, he claimed that it was a view so ethereal it must be like those seen by angels. In those moments by the Falls, I felt an angel by my side. I had been suffering an existential crisis ever since Kalpana's death, as I don't believe in life after death and yet I often felt her presence. I couldn't understand how this could be possible. I even had dreams where she appeared, very real, and spoke with me. In one of the more memorable dreams, I complained about her having died and she tried to comfort me, saying "you can build a bridge across tomorrow."

Standing on the bridge at Victoria Falls, with Kalpana's spirit by my side, I realized I had built that bridge. And I've been much more at peace since then, still not believing in god or heaven, but believing in angels.

We later visited a local village, where it was interesting to see the inside of the round huts, called rondavels, in which most Zambians live. I gained an appreciation for how clean a "dirt" yard can be. The dirt is compacted and swept perfectly clean. Many of the homes and yards were quite nice. I preferred not to have to bargain with the fellows selling their crafts, but as that seemed to be the appropriate, acceptable practice Jim graciously assumed that job. We bought some very attractive carvings and baskets, while I stood by saying helpful things like, "That's not enough. It should cost more than that!"

A helicopter flight over the Falls was a really inclusive way to see them in all their glory, since when you are on the ground you can only see a small part of the whole. As the Falls tumble into a gorge there really isn't any spot on the earth that can give the vantage point of the aircraft. It was truly beautiful. While airborne we flew over a small national park near the Falls where there are five white rhinos – and we were fortunate to see one. These animals were purchased from South Africa, where they had been raised in captivity as all the white rhinos in Zambia had been hunted to extinction. One of these purchased animals now living "in the wild," albeit in a small park but at least there are no fences like in a zoo, is pregnant. Her baby will be the first wild rhino born in Zambia since their absence, so is cause for great anticipation. Cross your fingers!

With this "sighting," although it's hard to accept a glimpse from a helicopter as being a true sighting of a wild animal, we had now seen three of the "Big Five": elephant, lion, rhino, leopard and buffalo. These are the "most dangerous" of the African animals, that is, the most dangerous to the people who used to shoot them. These days the most dangerous are the hippos, since ignorant tourists often get between them and the water that the hippo wants to get to. The tourists end up trampled. Apparently this is the most common cause of death to tourists - trampled by a vegetarian, rather than being eaten by a lion.

After two nights at the Livingstone we were driven in a Bushtracks van to the small airport for our flight to Lusaka. The Lusaka airport was really interesting, full of hustle and bustle, many people excitedly awaiting flights. We saw mostly black Africans but also small numbers of diverse peoples, including women in full burqas which were the first I'd ever seen. Everyone was well dressed, happy and smiling. As we had approached the airport in the air, however, we'd looked down and seen some pretty tough-looking ghettos. Sad but nothing we could really do anything about on this trip. I kept reminding myself that we were helping the wildlife by taking these safaris, and that I couldn't fix all the continent's problems. But it was still hard to see. So, conversely, it was almost a relief to see all the cheerful people in the airport. We had some interesting bureaucratic paperwork to fill out but otherwise as usual no difficulties making the connections. And we didn't even have to bother with actual airplane tickets; we never showed our passports to anyone. We were always met by someone with a clipboard and either they figured out who we were without asking, or they asked us to point to our names and that was sufficient. Pretty fun!

We hopped on to our flight to Mfuwe, a small town towards the north along the Luangwa River. At this little airport we were met by our next guide, Deb, who is one of the "best female guides in Zambia" according to our travel book. She's originally from England and a lovely person. We truly enjoyed our time with her. For this first drive out to the camp called "Kaingo" (which means "leopard") she had brought along Brasdon, the "spotter," who is the person responsible for holding and sweeping the bright light during night drives. We were very excited as this was our first game drive at night.

It took about 45 minutes over a paved road to reach the entrance of the South Luangwa National Park and we really enjoyed this part of the drive. The road was lined by the homes and buildings of a small village and there were people everywhere, all of whom seemed busy at this time of day preparing for dinner. Many people were riding on bicycles; women were carrying five gallon containers of water on their heads and even some with babies in slings at the same time; children were running around, all of them smiling at us, waving and shouting "hello!" The children walking home from school were smartly dressed in uniforms with crisp white shirts and ties. The rondavels looked clean and well-built; the yards were all nicely swept; the women outside cooking looked up and smiled as we passed. It is a truly prosperous area and we were pleased to be told that the village is flourishing due to the safari tourist business. These people grew the foods that we ate in our camps, and the guides and camp staff were also from the local area. It was very uplifting.

We entered the Park and I fell deeper in love with Africa! I loved the flora and fauna of this part of Zambia even more than the Okavango delta. The large Luangwa River runs through the Park, to eventually join up with the Zambezi, which leads to Victoria Falls. Many different areas exist along the river, from deep shaded ebony forests, large wooded areas with mopane and sausage trees, to open grasslands and areas with small bushes. We saw huge flocks of glossy starlings - quite beautiful long-tailed birds - and "love birds" that looked like bright green carpets when they were pecking on the ground. Several prides of lions, hundreds of hippos and elephants and of course impala beyond counting and huge herds of zebras and giraffes – all this we saw over the course of our four night stay.

On our drive to camp the first night we stopped often along the way and met some of the birds we'd already gotten to know - hadeda and sacred ibis, black-winged stilt, red and grey hornbills, Egyptian geese, guinea fowl, and fork-tailed drongo. We also saw a large number of new ones: African spoonbill, saddlebilled stork, and hammerkop. We stopped for sundowners along the river near a family of giraffes who munched contentedly while we sipped our Mosi beer. These giraffe, called the "Thornicroft," are unique to this area. Deb seemed proud, almost protective, of them. And the same with the zebra who are a special subset of zebras, and the small antelope that inhabit the valley called "puku" are seen nowhere else. We came to love the puku for their intelligence and their beauty.

As night fell, Brasdon got out the large spotlight and stood up in the front of the open-topped Land Rover. He spent the next two hours sweeping the light from side to side and our heads swiveled back and forth as if we were watching a tennis tournament. We saw nightjars and water dikkops and several members of the mongoose family (genets and white-tailed) as well as a few elephant shrews. We saw our first buffalo - number four of the Big Five for us. It was really interesting and exciting to be out in the bush at night! Whenever we came across a diurnal animal such as an elephant or antelope, Brasdon was very careful to shine the light away from their eyes. We drove quite slowly so as not to startle any animals.

Driving into Kaingo camp at night was almost festive, with lighted kerosene lanterns along the road and the welcoming sitenge aglow. A full table set for dinner, complete with candles, greeted us. I almost cried as Kerri, another of the guides, and Sadie, the camp manager, rushed up to greet us with open arms. In fact I guess I did cry! We were offered a drink (Amarula all around), and shown to the facilities behind the sitenge. The bathroom, or "loo" as usually bathrooms are referred to in Zambia, was an adorable grass hut with a spiral entrance tunnel and thus no door necessary; the sink was propped on a lovely piece of wood with carved stones for counter tops. Back in the sitenge the same decorative motif was used, while a huge section of a leadwood tree was a centerpiece under the towering thatched roof with another section of the tree trunk on its side as the bar. The walls were only a few feet high, the rest of the sides of the structure were open to the night sounds.

After a delicious four-course dinner with wine we were led to our honeymoon "chalet." Built with stucco walls, it had a rustic thatch roof and many screened windows, one of which reached to the ceiling. Everything was made with wood and stone, and the floor was covered with soft woven mats. Outside along a stone pathway was a tub for outdoor bathing, with candles and bath salts and fluffy towels hanging nearby. All of this right on the edge of the river. Incredible. We left the kerosene lantern burning out front and dropped off to sleep until the night sounds again awoke us with pleasure.

We stayed at this camp for four nights and it became my favorite. The guides were very professional and yet excited about their work. Many times they pulled out cameras and excitedly snapped photos at the same time we did. The owner is a nice and thoughtful man, laid back and yet attentive to every need. His parents were visiting at the time we were there and they were lovely people too. Every mealtime or game drive was "announced" by drumming on the huge ceremonial drums kept near the sitenge and we grew to love that sound.

There are only five cabins at Kaingo camp and as most of them were empty there were never too many guests. We often enjoyed our drives or walks with just the guide and no other guests, so were fortunate to receive quite individualized attention. It impressed us that Deb and Kerri (who is from Zimbabwe) were in the process of writing instruction manuals to be provided essentially free of cost for young aspiring safari guides, so that more Zambians can become guides for the national parks. The books that would be required for study are outlandishly expensive and not many Zambians can afford them. These are women who clearly care about the future and are trying to make a difference … which felt wonderful to me, of course. There was one other guide at the camp, Patrick, whom they had trained since he began as a spotter years ago. Deb and Kerri certainly did a good job, as he was really fun and knowledgeable.

There were many highlights of the game drives, not just the night drives but also the walking safaris we took. It was exciting to be on the ground near hippos, elephants, giraffe, and buffalo but on the other hand we could get closer in the Land Rovers.

I remember the first time we saw lions from the car and I became worried. "Why haven't you gotten out your gun?" I whispered to our guide from the corner of my mouth, trying not to move my head or let the lion know I existed. I had this idea that the moment the lions saw us they would leap into the open top of the jeep and eat us. But the guides explained that lions rarely attack humans at all and essentially never attack people in cars. They have never been hunted from cars, only from the ground, and so they are much more wary of people on the ground. So, we were able to drive quite close to lions and even had some walk right up next to our Land Rover without any real bother.

During one evening game drive at Kaingo a pair of lions walked right next to the jeep, mated, and then flopped down on the ground right next to us. What a nice experience for our sundowners that night! It was really interesting to see that in the case of these lions the male was really trying to be in charge, deciding when to mate and following the lioness everywhere, even tripping her once when she tried to walk away, perhaps because he was too tired to follow just then. Apparently, during the three days that a lioness is in estrus she mates with her male every 10 or 15 minutes! So it is no surprise that he was tired.

In the case of leopards, which we also saw mating, although not in broad daylight but by the red light of our Land Rover's lamp as they are nocturnal animals, it seemed to be the female who called the shots, walking in front of the male and backing into him until he complied. Quite fascinating! We also saw several leopards lying down quietly in the bush and breathing hard through their mouths - called "flehmen" - sucking air over a scent gland in the back of the nasal passage, by which they can evaluate scent markings.

During our walking safaris, accompanied by a guide and an armed game scout, we learned all about the spoor and dung of the local animals, even going so far as to check inside some of the dung. The different kinds of termites were fascinating to study. Deb referred to part of the walk as "scratch and sniff," as we checked out the flora too. We finally saw our first buffalos in the light of day (Jim and I kept saying "tatonka!" and pointing our fingers up in the air on the sides of our heads in imitation of a buffalo's horns, like in "Dances with Wolves"). They are majestic looking creatures although a little funny with the "50's hairdo" shape of their horns and their shiny chocolate brown coats.

We also loved the times where we just sat out in the middle of Africa, listening or watching. The guides at this camp really understood the importance of standing quietly still at times, which we appreciated.

Down by the river there was a "hippo blind," where we could sit hidden on the bank of the river and get really close to the hippos. We saw the oxpeckers hopping about on the backs of the hippos and a baby hippo nursing. Near the camp was also an elephant blind, a platform up in a tree near where the elephants would "usually" cross the river.

One day after lunch, just as we were making plans to go the elephant blind, at that precise moment a group of six elephants walked right into camp and hung out around the sitenge. We were "trapped" inside for several hours! It was one of our most memorable moments of the trip, with a bull and two large females, two juveniles and one baby all captivating to watch. I especially enjoyed the baby. The family of elephants had many interactions with each other which were endearing to watch. And they were clearly always aware of our presence. The juveniles would turn to look at us, and flare their ears to warn us to stay away. It was cute to see the baby brave this move too.

At one point during the adventure Deb was walking in another part of the camp when the bull elephant charged at her. She yelled so as to teach him not to do that to humans, so that the game guards won't think he's a threat to visitors to the National Park. The yelling disturbed the large female elephant which was closest to us at the time. She turned and began to charge towards us. As the sitenge is quite open it was very scary and we all backed way into the corner behind the bar. All, that is, except Jim, who kept filming! But she stopped short of the building, flaring her ears and making her opinion known.

Another elephant was browsing outside our window one night keeping us awake with very loud thrashing of the trees and munching and snapping. Jim got up to take a video of the escapade. Just the tiny little sound the camera made when he turned it on caused this elephant to whirl around and glare into our window. Her huge tusks were touching the screen as she poked around the outside of our tent with her trunk. I was frozen in fear and Jim looked like such a vulnerable little pink thing standing silently still in the dark in front of this looming bulk. After a time long enough for us to have been completely scared witless, she simply turned back to her eating. Guess she had good hearing!

One day we visited the lovely Tafika Camp run by John and Carol Coppinger, to take an ultralight flight over grasslands and the nearby mopane forests. What a wonderful experience and a great way to really see Africa. We could enjoy hippos and elephants and buffaloes, and they didn't appear to be bothered by the craft at all. The crocodiles, however, leapt immediately into the water when we flew overhead. I felt a bit of satisfaction as they don't have much fear in their lives yet cause fear in so many other animals' lives - including my own! We looked down on an eagle's nest to see a baby eagle. It was really divine.

The canoe ride back across the river was also an experience. As we stepped out of the canoe after crossing the river and looked back, we saw a hippo rise up out of the water JUST where we had been. In a tiny delicate little canoe. Hmmm. That could have been bad. Remember what I said about hippos being dangerous?

Once back to the sandy river bank we felt safe, until we noticed Kerri's calculating glances towards a couple of reclining lions nearby and our Land Rover about 200 meters distant. She seemed to be hurrying and I'm glad that she didn't tell us until we REACHED the Land Rover and hopped in, that she'd been trying to determine if we could close the distance to the vehicle before the lions reached us in the event they decided to try.

On our last afternoon/night game drive at Kaingo we visited some lionesses and their cubs at a buffalo kill, which we'd found by spotting a tree full of vultures in the distance and then rushing cross country towards it. The buffalo hadn't been munched on much, so it appeared that the females were waiting for the males to come and eat their share first, even though the lionesses had brought it down. Harrumph! The cubs were leaping and running around, climbing bushes, posing on various logs. Vultures would hop over towards the kill, getting closer and closer until a lioness would suddenly get up, run over, and chase them off, followed by a cub who appeared to be saying, "yeah! Take THAT!" The lioness would then lie back down in the grass on the other side of a tree, leaving the kill open for another slow vulture advance. I guess there were too many flies for the lionesses to protect it by lying close by.

We then drove over to the river for sundowners and watched two hippos fighting. It was shocking how violent they were with their huge teeth tearing at each other's faces and splashing blood all around! Wow. Truly "nature red in tooth and claw" this day. We also saw a tiny pearlspotted owl take down a mouse right next to us after their own ferocious battle. It was a wild evening. Later we stopped at an open spot along the river to enjoy the luminous full moon. It was completely red. I've never seen it looking so magnificent. Derek told us that the local Africans say there is "rabbit" in the moon rather than a man, and we could really see its outline. They call it "Kalulu."

It was difficult when the last day at this camp arrived. I cried saying goodbye to the superb people with whom we'd shared so much. Those late nights out on the Land Rover, listening for leopards, following them across the bush, watching lions at play and protecting a kill, finding beautiful new birds, learning the kinds of sage brush and plants that created the scents I loved (and discovering that one of my favorite odors was actually zebra urine), stalking hippos and giraffes, getting up before sunrise every morning to a new adventure. GORGEOUS sunrises and sunsets, red skies …

It was very hard to leave.

We were driven out of the Park by Deb and Kerri after tearful goodbyes to Derek, his parents, Sadie, Brasdon, and Patrick. Again we enjoyed the cheerful calls of the children of the villages before reaching the airstrip and our little plane. We flew back to Lusaka where we had time for a pit stop before the plane flew on to "Jeki International Airport," a dirt strip along the river in the Lower Zambezi National Park, which is also in Zambia on the border with Zimbabwe.

The usual Land Rover was waiting for us with a cooler full of beer and gin and tonics. The camp host Rob was a cheerful fellow from Australia and we had a nice drive out with him to the Sausage Tree Camp on the river. The sitenge, as had become customary, was inviting and open and had magnificent views. The little guest loo was elegantly decorated with a family of endearing wooden guinea fowl which we fell in love with. Jim wanted to get a family of guinea fowl of our own.

Our room was spectacular. In this camp we had a circular white-roofed tent with smooth painted concrete floors and an attached loo also made of the concrete, but with only low walls and no ceiling. Taking outdoor showers was marvelous! I really enjoyed the frogs that would jump in and with a huge tree overhead, again we could do birding while brushing teeth. The bed was scrumptious with a huge fluffy comforter, and each night hot water bottles were placed under the covers on both sides of the bed to make it even more inviting. Such luxury - although the first night as I sat down on the edge of the bed to take off my shoes I felt something small and warm under the covers and nearly died of fright. I'm so glad I didn't try to bash it to death before peeping under the covers! As at all the other camps, laundry was done each day and our clothes were delivered to us at night smelling of the African air and sunshine.

This camp wasn't quite as rugged as Kaingo, however it was most intent upon making our stay elegant and romantic in addition to the focus on wild Africa and the animals. We listened to hippos and buffalo and elephants all night (and even peed right next to elephants under the moonlight in our low-walled bathroom). As always the camp guides were incredibly knowledgeable, able to find animals and birds that we would never have seen on our own. One day, for example, while we were stopped for tea and biscuits, I shouted out that I saw a lion. We grabbed our binoculars while our guide Aubrey calmly stated, "Long-haired mongoose." Oops. I guess that's not even CLOSE to a lion.

One game drive was almost devoted to birds exclusively and it was just Jim and I with two guides. We saw a vast number of birds and learned more and more about their calls and their habits. For example, the "dead-battery" bird, the water dikkop, whose call starts out robust but then peters off, trailing down, down, down. We took many boat rides out on the expansive Zambezi River. Jim even fished for tiger fish and when I saw that they had a mouth full of vicious-looking TEETH, I grabbed a rod and caught (and released of course) two of my own. They didn't seem too helpless with those pointed scary teeth, and we were warned never to trail our hands in the water.

On a canoe ride down a small side channel one day, we got very close to hippos in the water and noted that they could waddle out and up the bank with incredible speed. We enjoyed many shore birds, especially the colossal goliath herons. From now on, great blue herons will look tiny to me. There were many weavers and their intricate nests hanging from the trees. On the river banks you could see the burrowing nests of the white-fronted bee-eaters who were flitting in and out, parading their brilliant colors. Other amazing birds that we could watch were the kingfishers, especially the pied and the brown-hooded. Such vivid colors!

One evening we took our sundowners amongst some 25 elephants who had come down to the river to drink. We saw them caress each other in greeting, and watched a baby who could hardly control his trunk. He kept getting it stuck in his mouth and would have to shake his head violently, or he'd manage to get it twirling in a circle when he tried to throw sand over his back. Then he'd throw himself down on the sand and roll around like a puppy. Another elephant used his head against the trunk of a tree to vigorously shake it and then was able to pick up all the seed pods that fell. We enjoyed the elephants for a long time, watching the family interactions. Remarkable to be so close to all of this!

Another evening the camp hosts placed a table complete with white tablecloth, umbrella, and chairs out in a shallow part of the river. Just before sundown we boated near to the spot, took off our shoes, and walked in a few inches of water across the sandy river bottom to sit at the table. Such a sweet idea! We chatted quietly as we drank our sundowners and watched a dazzling sunset over the Zambezi escarpment of the Rift. Our hosts explained that if crocodiles DID happen along, we'd see them before they would have a chance to get to us, and theoretically we'd have enough time to get back to the boat.

The funny part of this experience for me was that before we left for Africa I'd been warned that there is a danger of infections from parasites in the rivers and that under no circumstances was I to ever even CONSIDER letting any of this contaminated water touch my body. So when the boat pulled up near the table in the water and our host took off his shoes, climbed over the gunwale, and motioned for us to follow him through the shallow water, I balked. I realized, however, that I couldn't simply refuse to join in this romantic beautiful setting and so I tip-toed across the water as fast as I could and then pulled my feet up onto the seat of my chair as soon as I sat down. Rob noticed my discomfort and asked why my feet were elevated. I explained that I had a fear of putting my feet in the water as I didn't want to become infected.

"I see," he commented. "But, the water is perfectly safe; as a matter of fact you've been SHOWERING in this river water for DAYS!"

On another evening, the hosts secretly placed a table complete with flowers and candlelight, and surrounded by kerosene lanterns, out in front of our tent on the bank of the river so that we could dine alone. Our butler stood nearby, handing us our wine and food; the chef (who was from the Congo but educated in Paris) came by to explain the menu. We also had our own roaring fire where we sat for our pre-dinner cocktails and appetizers and watched the fireflies down by the water. Of course I cried! It was magical.

After three nights at this camp we had to say good bye to Africa. This was very hard. It's hard to even write about it. One thing I will sorely miss is the friendliness. In Zambia and Botswana, Africans greet each other each day with "Good morning! Have you slept well?" What a lovely touch.

We'll be back ... we're already thinking of seeing the red sand dunes in Namibia ... going for a walking safari in the North Luangwa Park ... and going back to Kaingo.

Yes!

Chapter 2
Africa 2004

✦

Magical, enchanting ... and unforgettable.

We started in Namibia, a wonderful country on the west coast of southern Africa. It is about twice the size of California but with only 1.5 million inhabitants. As most of the citizens live in the capital city, Windhoek (pronounced "ven-took"), the rest of the country is pretty wild and untouched. Just the way we like it!

Namibia was previously part of South Africa and was called "Southwest Africa." It gained its independence in 1990. The Namibian constitution states: "biological diversity ... and living natural resources are to be utilized on a sustainable basis for the benefit of all Namibians, both present and future." We were excited to be contributing to this country's ecology by visiting!

Our journey took us via London, where we had an 11 hour layover. We zipped into town and went to our "favorite" restaurant once again. This time, the owners recognized us. The crazy couple that always stops by on their way overseas (last summer on the way to Africa, and in January on the way to India). We had a delicious lunch, and then slept for a blissful three hours in a little hotel room that Jim had booked near Paddington Station before returning for our next 11 flight from London to Johannesburg, where we had a 2 hour wait and then a 3 hour flight to Windhoek.

The runway that we landed on was quite far from the city, in fact, we couldn't see any city at all, just desert. With a huge landing strip. We got off the plane and stood on the tarmac for a few moments, until a pilot came up to us and asked if we were the ones who were going to Little Kulala. What service! We pointed to our luggage, which was carefully laid out on the side of the runway, and then hopped in the Cessna 210, which was parked right next to the 737 we'd arrived on, with two other passengers.

The one hour flight to Little Kulala, our safari camp in the Namib Naukluft Park (near the southwest coast), was eventful. The view started out rather boring, just scrub desert and low mountains, but soon turned to indescribable beauty… expansive red dunes, sharp black rocky hills, and open grasslands of golden Bushman grass. I was beside myself with excitement; I hadn't been able to imagine such glorious beauty.

Jim, however, was having different feelings - those of nausea, which we figured was due to the very loud plane and no ear plugs, together with little air circulation and a very bumpy ride. So he was visiting with the inside of an airsickness bag, rather than looking out the window. He felt better when we landed on the tiny dirt strip and were met by our guide, Nigel, who drove us via jeep to our camp.

There we saw our first animal of the trip - a gorgeous and delicate springbok, running and leaping along the side of the landing strip. What a sweet antelope! Very thin legs, light beige tummy and darker brown back separated by a very dark black stripe along the flank. His delicate little horns pointed straight up, with a light yellow swirl also spiraling upwards.

Nigel stopped to point out a pair of Ruppel's koorhans, birds the size of quail who were strutting along amongst the gravel and rocks. Our first "life" bird of the trip!

We simply LOVED our safari camp, Little Kulala. It has 8 chalets and a main lodge, designed with a combination of African and Southwest style. The chalet had a front section that was a "typical" safari tent, with wooden flooring and a huge comfy bed with fluffy down comforter, while the back section was made of stucco. This small room was a very elegant marble and tile bathroom which opened out onto an outdoor rock garden surrounded by a high stucco wall and with a luxurious outdoor shower. Simply perfect. Imagine taking a shower outside in the desert air, listening to the sounds of birds, and gazing at the deep blue sky overhead. But there's more. Out in front of the tent was a wooden deck, looking out towards the red Sossusvlei dunes (the highest sand dunes in the world, up to 330 meters high) with a small plunge pool off to the side. A wooden ladder was built against the outside wall of the stucco bathroom leading up to the flat rooftop patio for sleeping out under the stars. What a fantastic place!

Moments later our first "sundowners" were poured for us on our rooftop. A delicious plate of appetizers was delivered with glasses of Amarula, and we watched the sun set on our first night back in Africa.

Walking on the sand trail to the main lodge, we enjoyed the tracks of the little animals and birds on the soft sand. The geckos started to call at this time of night and we enjoyed hearing their loud "barks" every few minutes.

We dined on the deck of the main lodge and were served course after course of gourmet food, washed down by delicious red wine. Ahhh … the safari life is a pampered one!

The hot water bottles in bed were a sumptuous treat. I got up several times at night to gaze out at the stars and the full moon. Deep quiet desert all around. Not a sound to be heard.

We were woken up at 5:15am with a "good morning!" and had a quick cup of tea and croissant before leaving for a jeep drive in the dark to go on a ballooning safari. I was amazed at the brightness of the stars, especially the star I took to be Sirius (to the left of Orion's belt). However, I had to do a double take. That wasn't Sirius … Orion's dagger was going UP from his belt, not down! I forgot that the constellations are inverted when you're in the southern hemisphere! So that incredibly bright "star" was actually Venus.

We got to the spot where the two balloons were being inflated and it was exciting to watch the proceedings. A large number of people were required to help with preparing the balloons, pulling out on sections of the bag as the air was blown in, and arranging all the ropes for the baskets. Then, we were off!

It was a fun flight, and a great way to welcome the sunrise. We flew over the desert for an hour or so and had fantastic views all around. We could even look down on our safari camp. When we landed, the gang was very careful not to mess up the desert floor (they even swept the ground afterwards) and "flew" the balloons onto the flat bed of the trucks. We were given a full champagne breakfast complete with a machete-flourish to open the champagne bottles. Jim enjoyed tasting all the new foods - zebra and ostrich and kudu - hmmm. I guess I forgive him.

Back at camp, we relaxed on the porch, watching a large group of sociable weavers coming and going from their nest over the waterhole, and the Namaqua sandgrouse coming up to drink. The males collect water in the feathers of their chests to bring back to the nests for the baby birds to drink.

After lunch, another fantastic gourmet meal, we went for a hike. It was the first time Jim and I had ever been out walking by ourselves in Africa. The reason we could do this here is that in the desert there are so few animals there's no chance of inadvertently coming across an elephant or a lion. We walked along a trail over to a nearby black rock mountain and climbed to the top. It was an interesting walk with lots of cool and unusual rocks in the ancient riverbeds. And it was romantic to be at the top of the mountain looking out over the desert. We saw a few new birds, including a lesser grey shrike, a familiar chat, and an acacia pied barbet with its gorgeous black and white striped head, red spot over the eyes, and brilliant sparkly yellow spots all down the wings.

We then went on an afternoon "nature drive" ... note I call it a nature drive and not a "game drive," as there were so few animals. We got out to see a massive sociable weaver's nest, and enjoyed watching a large spotted eagle owl, first perched and swiveling his head all the way around, then flying away with slow ponderous wing beats.

We saw our first gemsbok oryx – what a thrill! Such a GORGEOUS animal, majestic and magical, with striking black and white markings on the face, long slender straight horns pointing skywards, and a grayish-brown body the size of a horse. We've wanted to see one ever since we fell in love with a painting of one that we saw in South Africa last year, a painting which we eventually bought even though we'd never actually seen the animal. So now we'd seen our first gemsbok (pronounced "hemsbok").

Our drive continued up to the foot of the mountains where we had our sundowners on the enormous spreading alluvial fan covered with Bushman grass. We enjoyed a spectacular sunset, and passed a fun time chatting with Nigel and the other two guests, Danny and Claire. We saw our first ostrich – there were quite a few of them wandering around the grasses and munching.

We sat and reveled in watching the waving golden grass ...

Dinner was again simply incredible and I learned that I do like hake, the only other fish besides salmon that I've ever liked. We thought it was funny that the red wine we were served was chilled. And not only chilled, but actually FROZEN. We tried warming it up with our hands, and then by holding it over the candle ...but nothing worked. It was still ice cold by the end of the meal, which elicited plenty of giggling.

We tried sleeping out on our roof, but even though they had put out a comfy down sleeping bag, inside a canvas cover, we were simply too cold to stay for very long. The desert certainly gets chilly at night. Also, the moon came up and suddenly it was as if we were trying to sleep under a bright light. So we retired to our room below, and our hot water bottles.

Our next day's adventure was a trip to the dunes of Sossusvlei (pronounced "sossus-flay"). We were up again at 5:15am, and at the dunes in time for sunrise. A photographer's dream! The dunes are swirled in fantastic shapes, and the sunlight shining on one side – showing the deep red color – gives a sharp black outline against the dark, unlighted, side of the dune. Stunning.

We drove down a river course between the dunes and stopped along the way, enjoying each of the gorgeous formations. There was one dune that we were able to climb (called "Dune 45," as they are numbered), that was 80 meters high. We clambered to the top with Danny and Claire, and then ran down the side, which was so playful, as the sand is very soft and SO red!

We hiked over the sand dunes to a beautiful remote vlei, or pan, where water once collected. This one, called "Dead Vlei," was covered with chalky white clay and stark dead trees, amazingly picturesque. I missed Jim on the hike – he'd stayed back as he wasn't feeling great, and knew that he would have loved the scenery. By the time we returned to the vehicle, he was much worse, so we rushed back to camp, sighting springing springbok and running ostrich along the way, and doctored Jim with anti-nausea pills and lots of rehydrating fluid. I tucked him in bed and hoped for the best. We decided that the reason he was sick this time must have been all the food he was eating, especially all the new and strange kinds of meat. Perhaps his stomach was simply not up to the task.

It was sad to have my Amarula sundowner by myself and I began to miss our kids. I enjoyed watching the African pied crows at the waterhole, though, with their pretty white tuxedo vests. Danny and Claire invited me to have dinner with them, and we had a pleasant time although I seriously missed Jim. After dinner, the kitchen staff came out to sing to us. It was incredible. The songs sounded like those on Paul Simon's Graceland album, and were amazingly moving. One song was in English while the rest were sung in one of the African "click" languages (whereby some of the words include a click of the tongue) -- so exotic sounding. The young exuberant staff stomped their feet and marched around the room, clapping hands and laughing and smiling. It was wonderful, and more so because Jim could hear it from our room. I was glad he didn't miss it completely.

By morning I was relieved that Jim was feeling better, I got up at the usual 5:15am and took photos of the dead trees against the sky and Venus, and enjoyed being out in the quiet desert in the dark. We were taken to the dirt air strip for the trip to our next camp. It was hard to say goodbye to our sweet chalet, and the desert sand, and the red dunes and the Bushman grass. We both want to come back.

At the airstrip, we had to wait for about five minutes before our plane arrived. It is simply incredible how impeccably all of our travel had been arranged. The plane dropped off some kids who were coming from their villages to work at Little Kulala for the first time. They were very excited as it had been their first time up in an airplane, and some of them had been scared. They seemed very happy to be on the ground.

Our female pilot (yeah!) then flew us in a Cessna 210 to Swakopmund, a small town on the coast a little to the north. We flew over the red sand dunes, miles and miles and miles of dunes, for most of the one hour flight, and we flew low, only a few hundred feet above the sand. It was exactly like the scene in "The English Patient"! Very romantic and beautiful, and Jim didn't get sick! And then: our first view of the African coast. Wow.

We had five minutes to stop in the bathroom at the airport, and then hopped into a larger plane, a 14-seater Grand Caravan. There were just four other passengers. The pilot explained that it was to be a sight-seeing trip up the coast to the northern part of the Namib desert, called the Skeleton Coast due to some old shipwrecks left along the shore and whale bones scattered on the beaches. This is the most desolate and untouched part of this hauntingly mysterious desert, with sweeping vistas of wind-swept dunes, and dense coastal fog caused by the cold Benguela ocean current running directly up from Antarctica. Mightiness, solitude, and COLD.

The three hour flight took us over a stunning landscape. We flew low again, over huge seal colonies, little tiny "holiday towns" and campsites on the ocean, and down into massive river channels in the mountains along the coast. After flying over a seemingly endless plain of grey gravel scree and twisting yellow dunes, we landed at the "airstrip." Actually, it was just a section of gravel that was outlined with stone markers. The jeep picked us up, and drove us the 200 meters to the camp, nestled down in the sand of a dry river valley.

The camp had 7 tents, and we were given the "honeymoon suite"! This was a typical safari tent, with raised wooden floor and again the delicious huge bed with fluffy comforter. Just like at Kulala, the front of the tent was a set of glass doors, instead of the old fashioned zippered front. This is very helpful when it's cold or windy as you can have it all closed up and yet still look out at the view. And the view out of our tent was wonderful ... sand dunes with a few small Euphorbia bushes, and a small rocky outcrop of grey and pink granite. The bathroom had a nice open slate tiled shower and a sink in a barrel fitting with the nautical theme.

The Skeleton Coast is very remote, and is an especially delicate ecosystem. The camp, the only one in the reserve, is very conscious of the need to protect the environment. Water is brought to the camp by a tractor and so is in very limited supply. We collected the cold water from the shower tap (before the hot water arrived) in a metal pail, to be used later. And the only electricity was from solar power. We took our showers in the evenings as that's the only time the water was really warm!

We all sat down to lunch and got to know each other and our hosts. The camp is run by Douw (pronounced "doe") and Dahleene, with the help of Uys (pronounced "Ace"). They are all Namibians from Windhoek who were raised speaking Afrikaans and English. They also know a smattering of "click" languages, which they use to communicate with the staff, all of whom were of native descent. Dahleene was the organizer of the camp, and the two guys were our guides for our three days of adventure. They were incredibly sweet and brilliantly informed about the geology, animals, plants, and ecology of the region. We were shocked to find out later that they were only 23 and 24 years old; they seemed to be in their 30's in terms of their maturity. But they loved joking. Jim said they were like the Blues Brothers, sitting in the front of the jeep, wearing dark glasses, telling jokes and horsing around.

The other guests were a young couple from Italy (Chicco and Ludovicka), and female friends from New York, Joan and Cathy.

We started with a delicious multiple-course lunch, entertained by a little dusky sunbird who was hopping about the bushes right in the window of the main room and who would let us walk right up to him. The afternoon nature walk was a GREAT deal of fun. We learned an amazing amount about the area and the few insects and creatures we saw included several kinds of geckos, one of which we could pick up as he was so cold and immovable. I found a porcupine quill and put it in the brim of my hat. And there were many kinds of lichen. Douw and Uys were very careful about our not walking on the lichen, or on parts of the scree that would not regenerate. It was freezing cold and windy, however, and when we came up out of the riverbed onto the plain, it was incredibly barren and stark with cloudy overcast. Surrounded by rocky hills of streaked black, grey, and pink granite, we were reminded of Iceland! We climbed several of the hills, and enjoyed the expansive views.

As we walked back towards the camp we came across a basket of sundowners that had been brought up the hillside for us. Such a sweet touch. We all tried to down a beer (the Namibian beer, Tagel, is quite delicious) but it was so cold that we bolted them and ran to our tents to warm up with showers.

Toasty again, we walked over the sand to the main tent for dinner, along a path outlined by kerosene lanterns. So romantic! Along the way we saw two jackals, the first we've ever seen. They were so cute, like little dogs or foxes. A fantastic dinner was served; the menu was announced to us first by the waiters, with Isaac speaking in English followed by Yvonne in her "click" language.

By the time we got back to our room, we realized that we were going to have to increase the amount of clothing we wore at night. At Little Kulala, we were happy with T-shirts and the comforter and hot water bottles. But here, so close to the coast, with the marine layer moving in, we put on our full sets of long underwear and socks. Later, we added the bathrobes that they provided in the rooms – even with the hot water bottles and comforters, and a second set of blankets on top! I don't think I've ever stayed anywhere that was as cold as this coastline.

Jim had a miserable night, becoming quite sick. I was rather chilly running around at 2 am, emptying the pail - what a doting wife! We were actually rather scared by his being sick again, and wondered if it could still be from the quantity or type of food, or if it was just that he had started eating rich food too soon after being sick before, or whether this time he was really ill with a dread disease. I pumped him full of drugs: ibuprofen, Imodium, Pepto Bismol, and finally Ciprofloxacin. I wanted to make sure we covered all bases. In the morning, he explained that he didn't feel as if he could get up, but that I should go for the day's activities and take lots of photos. So I did, but it was hard to leave him especially as I still didn't know what was ailing him. I asked Dahleene to look in on him periodically and to bring him toast, rice and rehydrating fluid.

So, I went without Jim but thought about him and talked about him all day, which helped make up for his absence. Our adventure began with the usual wake-up at 6:30am, with a tray of coffee and tea delivered to the room, which helped get me out of bed in the freezing cold. Over at the main room a delicious full breakfast was served, and then we were off for our safari drive. This camp uses a closed vehicle as the weather is so extreme, and we were all quite happy to be inside and not out in the breeze.

We stopped every few moments, and enjoyed our first family of giraffe. So regal and dainty as they sauntered tall. As we drove inland we were surrounded by towering cliffs and mesas and buttes that look exactly like the American southwest. I had no idea that it was going to look identical to Utah but with giraffe and oryx! Incredible and beautiful.

We stopped for coffee and cake under a very old leadwood tree - thousands of years old apparently - and enjoyed lots of birds: white-backed mousebird, pririt batis, chestnut-vented tit babbler, swallow-tailed bee-eater, grey hornbill, mountain wheatear, and rock pigeons.

We came across a summer settlement of Himba, which was deserted being that it was now winter. The Himba are one of the oldest tribes in Africa; only the Bushman and Hottentot are considered to be more indigenous. They are a cattle-herding nomadic people and still live in villages in their traditional fashion, even though they are completely aware of modern 21st century habits - they just prefer their way of life! They do, however, like to get a little money now and then for purchasing metal or beer, and so they like to have visitors come and buy handicrafts. They don't speak any English and so there is no "hard sell" ... no bargaining or pushing of their goods or anything like that. They just put out their work for display, and if you want to buy, you can.

We enjoyed seeing the deserted settlement as we could go inside the houses and take lots of photos. This village had five dwellings, four of them round huts and one square. Each one was small and low, with room inside for a fire and a few people. Pots and other implements hung overhead. The huts were built of sticks and covered with mud and dung, and arranged in a circle around the central "kraal," the cattle corral made of thorn bushes to protect the cattle. The ground was hard packed and swept very clean. A simple lifestyle that seemed pleasant actually!

We then drove on a ways to their winter settlement and enjoyed visiting the 15 or 20 people who live there. The men were all away tending to the cattle so we only saw the women, children and the old chief. They still dress in their traditional style, which is simply a small hide skirt with lots of beads and jewelry. But nothing else! They rub their skin with ochre (red clay), so it becomes a lovely shade of red. They also rub the ochre in their hair and make long red dreadlocks. What a pretty people! The babies were SO adorable that I wanted to grab them and cuddle them up. They wear nothing but a string of beads around the tummy and the ankles. They would tumble and run around the gravel, looking at us sideways and giggling at us, and then would jump back in their mothers' laps to nurse.

I bought necklaces for Jim and me, made of ostrich egg shells alternating with melon seeds. The seeds in Jim's necklace were from the Tsama melon and in mine, from the !Nara melon (the "!" infers a click).

After the visit we drove down to the river bed, and had a wonderful lunch. A table was set up attached to the side of the vehicle and chairs were placed out for everyone. A large number of different dishes, including lots of vegetarian choices, were set out along with beer and water and soft drinks. Plenty of birds lunched with us: the grey lourie, guinea fowl, the brilliant crimson-breasted shrike, red-billed Francolin, pale-cheeked babbler, Egyptian goose, three-banded plover, blacksmith plover, rock martin, and palm swift.

One of our funniest moments was when Uys stopped the car for a moment so that Douw could discuss some aspects of the geology with us. Of course, I was quite interested, as it was fun to understand how the various red buttes and mesas were formed, and why there were tall mountains of pink and grey stripes. As he was talking, however, several of us were glancing about with our binoculars and a comedy in 4 parts ensued, with all four of us making these comments in such quick succession that later, it seemed we were all talking at once:

Douw (looking back at us from the left-hand front passenger seat as driver is on the right in this country):

"So, as I was saying, there was a blah blah tectonic blah blah blah uplifting blah blah" (I can't remember exactly WHAT he was pontificating about at the time, except that it had something to do with the geology of the region)."

Cynthia (excitedly, while holding up binoculars, from the left back seat):
"OH! What bird is that?"
Douw:
"Could you just let me finish this one story first?"
Chicco (with curiosity, while holding up binoculars, from the right back seat):
"Is that a DOG?"

Douw:

"If you could just let me finish this story …"

Uys (from the right front seat, looking in direction of Chicco's binoculars):

"Oh SHIT, oh SHIT, oh SHIT!"

And I swung my binoculars over, just in time to see TWO CHEETAHS run in front of our vehicle! They were sleek and clean and gorgeous. Incredibly powerful. Stunning. Moreover, they are hardly EVER spotted. Douw and Uys had never seen one in the Skeleton Coast in the entire one and half years they'd been taking nature drives every day. So that explains Uys' lack of courtesy with his language … and the rest of us were thrilled and talked and laughed about it all day. "Could you just let me finish …?" became a recurring theme.

As soon as the cheetahs passed out of sight behind some rocks we zoomed and bumped along the road to try to see where they'd gone but to no avail, we didn't see them again.

It was quite an excitement.

The next activity was an "elephant hunt," down amongst the trees and bushes of the dry bed of the Hourasib River. The elephants in this region are "desert adapted," and have made many changes to their lifestyle - and even their biochemistry - to be able to live in such harsh conditions. There are very few of them, and hard to find. We realized we had some serious tracking to do, and so three of us, Chicco, Ludovicka, and I, climbed up to seats on the roof of the vehicle to help with the search.

We drove up and down the river, around and around various bushes and trees, following elephant tracks. Uys and Douw kept pointing them out to us, and explaining which tracks were fresh and which were old, but WE couldn't even tell which direction the tracks were pointed. We were certain our guides were going the wrong way but figured that perhaps they knew more than we. At one point, Chicco and I shouted, "STOP!" exclaiming, "Fresh elephant dung!" Uys put the jeep in reverse and scrambled backwards over the sand, churning up dust in our faces, only to look at the purported fresh elephant dung and state, "Well, that's about 2 months old."

We were a little embarrassed. Uys confirmed, "It's only fresh if it's steaming and stinking!" Well, we never saw any of THAT.

After an exceedingly long time of driving and searching and scanning the trees and deep thickets, Joan suddenly pointed and yelled "Elephant!" Of course, we were supposed to be QUIET, but luckily the elephant in question didn't rush off so we got to see desert elephants - three of them, two female adults and a young adolescent. Lovely. Powerful and yet sweet, as we drove up rather close and they gazed out at us from those deep, expressive eyes. We watched them for a long time while they browsed through the bushes, and then wandered down the river to a new spot. Unforgettable.

It was getting late in the day and with a long trek back to camp, we set off. The drive continued across long expanses of grey scree and as the sun set we saw numerous large groups of ostriches running across the plains in huge numbers. There were about 15 groups of them, each with about 20 ostriches, and they seemed to be running as fast as they could - 50 or 60 km/hour! They twirled and danced and jostled their bustles as they went, fluffy black and white "feather boas" swinging around in the air. A breathtaking, evocative sight.

Douw stopped the car at a lovely open spot out on the steppe, as the Southern Cross rose in the sky, and he read us a poem, "Wilderness" by Ian McCullum:

> Have we forgotten
> that wilderness is not a place,
> but a pattern of the soul
> where every tree, every bird and beast
> is a soul maker?
>
> Have we forgotten
> that wilderness is not a place
> but a moving feast of the stars,
> footprints, scales, and beginnings?
>
> Since when
> did we become afraid of the night
> and that only the bright stars count?
> or that our moon is not a moon
> unless it is full?
>
> By whose command
> were the animals
> through groping fingers,
> one for each hand,
> reduced to the big and little five?

Have we forgotten
that every creature is within us
carried by tides
of earthly blood
and that we named them?

Have we forgotten
that wilderness is not a place
but a season
and that we are in its
final hour.

A magnificent poem that brought chills to my skin.

When we returned to camp, Jim was up from bed and walking about. As he met us at the car I simply melted in his arms with relief that he was well. From this point on, he was MUCH more careful with his food intake, starting with toast and white rice and slowly increasing to richer foods over the next few days. Guess he should have done that after the first time he got sick.

He'd had a rather uneventful day, mostly sleeping, but he did have a nice encounter with an oryx by our tent that allowed him to come quite close for photos. The locals call him "Prince Albert," and he seems to like to hang around the camp.

Dinner was lovely again, and sleep was perfect, albeit cold. It was terrific to go outside at 2am and see billions of stars, and feel the quiet.

The next morning we were up at 6am, and Jim was able to join us on the day's activities - yeah! After a quick breakfast we drove across the desert, which may as well have been Mars for its barrenness, to a more southern section of the Hourasib River - a fantastic oasis, with palm trees set amongst desert mountains of red rock and sand. We stopped for our coffee break in a part of the sandy river bed that was quite wide, with towering rocks of fantastic shapes on one side that we enjoyed looking at.

While climbing behind some bushes for a relief break, Cathy said that she heard a LION! We all ran to the jeep and huddled there a bit frightened, but Uys and Douw investigated and didn't find any tracks.

We traveled on down the river for a while until we came to a spot that had a canyon angling off to one side. We got out for a walk down the canyon. The guys promised that no lions would go down this canyon as it's only dry sand and rocks with no food. I thought to myself, "Except us..." The canyon had lots of tall clay formations that had been created by weathering, and was fascinating. We learned about Bushman grass seed, and how it has a little "flag" sticking out of the seed case that allows the seed to dig itself down into the sand when the wind blows.

At the end of the canyon Douw asked us all to sit for a moment of silence. SO quiet. Very moving. I was reminded of my friend Kalpana, who loved to sit and FEEL the silence. I cried a bit.

We returned to the jeep and started off on our lion hunt. The guys found some fresh tracks and we followed them for a few miles. At one point we stopped for Jim to take a picture of an oryx against the sand dunes. Everyone seemed to get into the action while Uys explained that the oryx we were looking at was going to turn his head in a few seconds and walk right in front of the dune. In a few seconds the oryx did exactly that, which was great.

We saw several new birds along the way: Cape wagtail, auger buzzard, rock kestrel, common waxbill, bokmakerie (bright yellow, green, and grey), tractrac chat, and black crow.

We saw plenty of lion tracks, some of them were massive and we followed them INTO a deep thicket on one side of the river bed, but didn't find any leading OUT. So, we were pretty sure the lions were IN the thicket. Too bad they didn't show themselves.

Instead we ventured down another side canyon, this one wide enough that we could drive through, although it was quite wet. The ground still had a lot of water remaining from the last wet season. We hopped out and Uys and Douw showed us that along the sides of the dry river bed, up against the lofty black and pink walls of the canyon, there was quick sand! It was remarkable to see how fast our bodies would sink down but luckily, only to the tops of our thighs, not all the way down. Jim was the first to try it ... he zipped off his pant legs and took off his shoes and waded in ... and then the rest of us got brave and tried too. It was a lot of fun!

We had an entertaining time on one of the expansive sand dunes, clambering to the top. It was quite hard to ascend, as we were scrambling straight up the face of the dune and different from when I'd climbed the dune at Sossusvlei, where I had been maneuvering along a ridge line where the sand was hard packed.

At the top, you could see forever ... across miles and miles of dunes, all the way to the sea! Such a sight, like something out of Lawrence of Arabia, which is how we felt when we came back from our hike around the top of the dunes and skied down the face. We figured we looked like conquering heroes coming out of the desert, momentarily enjoying our delusions of grandeur.

The guys served up another great lunch alongside the jeep. The cold beer was delicious after all that blistering hot sand hiking. We then drove all the way out to the coast where the scenery became really desolate and barren, just dunes of sand and nothing else, as far as you could see. The wind was howling, and the sand was blowing. Douw referred to it as "Highway to hell." All of a sudden, he turned the jeep TOWARDS one of the dunes, and began driving UP on it! I was frightened out of my wits but everyone else seemed to love it. As we drove and drove, faster and faster, the wind was whipping the sand around, and it appeared that we were driving on clouds since the surface had a layer of moving sand. "Highway to heaven!" shouted Joan. It was other-worldly.

When we reached the top, the sand was hard packed enough that they could stop the jeep and we got out. Uys and Douw led us over to the crest of one particular concave dune, facing down wind, and had everyone sit in a row along the top (squinting and desperately trying not to get sand in our eyes). Then, on the count of three, we were all to start sliding down the dune on our butts.

And what a SOUND! The dunes actually ROARED as we moved. We later learned the scientific explanation, that there was a lot of air trapped on the lee side of the dune, which we were compressing out as we moved. It was very amusing.

Then, back down the dunes and out onto the coast. We drove back to the camp all along the seashore, where the yellow sand dunes "march" along the grey scree, being blown by the wind, and changing shapes even as we watched. We stopped to collect agates, and to pay tribute to a memorial for one of the "skeleton" ships off the coast and the crewmen who died. Douw told us the miraculous stories of the heroism involved in saving many of the passengers, landed on this harsh inaccessible coast. More bird life enthralled us - we saw African black oystercatchers, white-breasted cormorants, and kelp gulls.

Right before we got back to camp, we saw a group of springbok PRONKING! A "pronk" is when the animal leaps straight up into the air, with head and legs pointed downwards, jolting up and down several times while moving away. It's really adorable to watch.

After a luxurious shower, we did some star gazing with our telescope and saw various star clusters, then Jupiter and four of his moons. It was time for our "brie" (pronounced "bry," not "bree" like the cheese), or barbeque. It was too cold to eat outside, but everything was cooked on the outdoor fire and then brought in - it was great. And then, the kitchen staff came in and sang to us.

There were 6 or 7 of them, youthful kids in their 20's, and they were all natives, speaking the "click" languages. They were smiling and laughing, with a continual banter about which song to sing next. It was a moment to be cherished.

After listening to several lovely songs, Uys, Dahleene, and Douw joined them with a song in Afrikaans. The most entertaining part was when they did a silly song while holding a spear, called "the lion hunt" and they managed to get each of us up from the table, and into the act. Soon, all of us were cavorting around the room, yelling and grunting and flailing the spear. Quite an uproar!

After dinner, we all sat by the fire outside and reflected on the day and listened to the crickets chirping. Both Douw and Uys told stories, and then Uys got up and recited the beginning verse of the poem, "Auguries of Innocence," by William Blake:

To see a world
    in a grain of sand
And a heaven
    in a wild flower
Hold infinity
    in the palm of your hand
And eternity
    in an hour.

Filled with spiritual feelings, we all went off to bed … and heard jackals and hyenas during the night.

The next morning, after our coffee and tea were delivered to the room, and we had our breakfast, the plane arrived and we were off to our next camp. It was hard to say good bye, and as we took off from the gravel Uys and Douw and all the staff were waving and waving. Such a hauntingly beautiful, spiritual place. It was hard to leave as we'd become friends.

We flew in the Caravan - with another female pilot - for one and half hours, landing on the dirt strip at the private game park, "Ongava." Ongava, which means "rhino" in Herero, one of the older tribal languages, is situated in the center of Namibia. It's adjacent to Etosha National Park, one of the world's largest parks, about the size of Switzerland. We were met by the ubiquitous jeep, and driven 30 minutes to our safari camp, "Little Ongava."

First, we met our guide, Sunday, and had lunch on the deck of the main lodge located up on a hill overlooking mopane woodland and a water hole. And then we were taken to our chalet.

Oh, wow. We had NEVER in our wildest dreams realized that we were going to stay at a place that was this lavish, in fact, we had thought it was just going to be another little elegant safari tent, like the others. Well, we were in for a shock.

There are three chalets perched on the hill, and to reach them a beautiful wooden walkway meanders up the hill amongst large boulders and trees. When we arrived at our chalet, we couldn't believe that it was just OUR room and not another lodge! It was a beautiful wooden structure, accented with rocks and decorated with tent material and thatch. The door opened into the "living room" the size of our living room at HOME. Next to that was the bedroom with a king-sized bed and French doors leading out to the deck, and next to that room was a bathroom that would have made Louis

IV proud. Huge double tub, stylish double sinks, slate tiled shower, all surrounded by gigantic plate glass windows looking out over the wooded park, with the loo and bidet in a little room behind. There was a long hallway with closets and shelves and a writing desk. Then, out in front of all this was a deck cantilevered over the hillside, with an infinity-style plunge pool tiled with slate, a fancy outdoor shower, and even an outdoor king-sized bed with a sturdy thatch roof for relaxing in the afternoon with a glass of champagne.

Well, it was sumptuous to the point of decadence, and we did enjoy it. The only problem was that the OTHER clientele who would choose to visit Africa in places like this were not exactly the kind of people we like. I won't be too negative here, except to say that we won't be staying at a place like this again. It was just too irritating to be with people who don't care a bit about the animals or the ecology, and wear high heels and gold lamé with plunging necklines after a day of game driving.

On the upside, there were dozens of little birds in the trees outside our room, sipping at the water in the plunge pool. Lesser-masked weavers, short-toed rock thrushes, white-browed sparrow-weavers, and cinnamon-breasted rock buntings. And one fork-tailed drongo sitting alone on a tree limb. It crossed my mind that the last time we saw a fork-tailed drongo, we were in India …

Down near the main lodge, we saw hyrax (also called rock dassies) which are adorable little guinea-pig-like creatures closely related to elephants, incredible as that may seem based on their size difference.

We went for an afternoon game drive throughout Ongava with Sunday, and saw lots of birds, including our first lilac-breasted roller of the trip or, ABBR as it is affectionately called: "another bloody beautiful roller." We saw cape glossy starling, white helmet-shrike, red-crested koorhan, Monteiro hornbill - only seen in this part of Namibia, and double-banded sandgrouse who always come to drink 10 minutes after sunset. We saw back-faced impala, which are endangered and only seen in this part of Namibia, lots of springbok, and one blue wildebeest. We also saw a steenbok, the TINIEST antelope (actually, the dik-dik is smaller, but we never saw one). So delicate, with petite little horns and titanic ears, very cute. Now I know why steenboks were Laurens van der Post's favorite antelopes. And, we saw an eland, such a magical, wondrous beast revered by the Bushman … who love the oryx, but love the eland more. We even saw a group of oryx babies, who are brown instead of grey, and look like little cows with Pan-like horns. And our first – very majestic - kudu antelope.

The best part of that game drive was the lions. We saw them right at sunset, walking to a water hole. There were two males and a female, and they walked right alongside our jeep. We saw one of the males flehmen (opening the mouth wide, to "taste" smells), and heard the female calling to her cubs.

We had our Amarula sundowners in the jeep, watching the lions as they relaxed by the waterhole.

Sunday used the spotlight on the way back to camp, and we saw a scrub hare. Then, we enjoyed a lovely bath in our tub, with starlight all around, and a delicious dinner in the lodge. On the way back to our chalet we saw a preying mantis (godlike, to the Bushman) and a huge stick insect which Jim tried to pick up, but he scrambled away with surprising agility. Sleeping in our bed was luxurious although I couldn't even be sure that Jim was THERE, the bed was so colossal.

The next day, after our wake up call at 6:30am and a full breakfast, we went on a game drive into Etosha National Park. There were just the two of us with Sunday so we got great personal attention. We saw 24 new species of birds, including the bateleur eagle which I decided may be my new favorite bird, it is SO beautiful. We got to see it up close, sitting on a kill, and also soaring overhead, titling like a tight-rope walker, which is what the name means. We also saw the shaft-tailed whyda with an amazingly long tail, and the secretary bird! What a crazy bird, walking along as if it were a PERSON! It is so much larger than we imagined, and prissy, as he picked his way along the ground. We also got to watch him fly and land, and in those motions he was graceful.

We visited several waterholes, and saw astonishing numbers of game, probably more animals than we've seen in Africa before. We were able to sit right next to the waterholes and in the one near the area called Okakuejo, we actually got out of the jeep and could sit on benches just outside a little rock wall alongside the waterhole! The animals feel quite safe there and are accustomed to having people watch them. It's stunning to see large groups of animals arrive, single file, take a drink, and then file out as another group of animals arrives. Herds of zebra, wildebeest, impala, springbok, kudu, and oryx, all taking turns. And there were ostrich and giraffe in the background, and black-backed jackals running through the Bushman grass.

And elephant! We went to a waterhole that the elephant prefer, and watched a large group of them, drinking and splashing and playing. Some were touching each other's trunks with immense joy, communicating and showing great affection. We watched them for a long time. I do love them, and especially the babies. It was comical to watch them trying to use their trunks and sometimes failing, so that the trunks would spin around their heads uselessly. The babies would scramble along after their parents, trying to keep up.

After a delicious lunch at our lodge, we went back to our room to relax. We took a plunge in our pool - ha! We only went in to our knees, as it was freezing, but then took a great bath and an outdoor shower. We listened to the songs of the weavers and buntings, and the swishing sounds of the flocks as they flew from the pool to the trees and back. While having our afternoon tea, we saw a black rhino also enjoying an afternoon pick-me-up at the waterhole down below our lodge.

The afternoon game drive took us through Ongava on a quest to find some other rhinos - they also have some white rhino here. Sunday asked if we had any other "requests" for animals to find, and we mentioned dik-diks and bushbabies. We didn't actually find any rhino, but had a fun time trying. We even got out of the jeep and did some tracking on foot with Sunday and his loaded rifle. That was pretty exciting as we saw a lot of huge lion and rhino tracks but still no lions or rhino. We saw signs of rutting, and scat, and broken branches, and felt that we were right on their trail. But no luck. We did see a few duiker - sweet little antelope. To our amazement, driving back to camp in the dark, with Sunday swinging the spotlight as he drove, he actually found a BUSHBABY clinging to a tree! I've always wanted to see one, and it was every bit as adorable as I imagined. I wanted to take it home.

We had a delectable gourmet dinner after being met with a tray of drinks when we arrived back at the lodge. Sunday expostulated about how important our visiting game parks was for conservation. We are very glad for that. He is involved in some of the work that Paul Newman does and gave us some literature to read about "Children in the Wilderness" (www.childreninthewilderness.com) which made us cry; this group is something we may want to get involved in.

The next day we had another 6:30am wake up, breakfast, and then another game drive into Etosha. This time we saw an African wildcat, also a rare sighting but which just looked like a house cat. We enjoyed seeing a large monitor lizard basking in the sun. Lots of ground and tree squirrels scuttled around and we saw several giraffe, one of which did a flehmen with his pendulous lips. We also drove to see the Etosha pan which is a vast 40 by 70 mile expanse of white clay where five million years ago there was a lake. A lone ostrich was running along that enormous immensity of nothing. Dotted around were umbrella acacia trees, and other acacias with sweet smelling yellow fruits that the giraffes were eating.

At one waterhole, we saw our first hartebeest, a very pretty and the fastest antelope. And we saw our first warthog - very cute animals as they charge about with their little short legs. Jim says they look like little boys without pants. We also saw a Kori bustard, Africa's largest flying bird, at this waterhole, so a good score all around!

We returned to the elephant's waterhole and spent a long time watching elephant herds arriving. Eventually there were about 70 elephant there! Wonderful ... the babies were running and plunging in, lying down in the water, covering themselves with mud. I was enchanted and especially loved to see them greet each other, intertwining their trunks.

At another waterhole out on the open plain, there were close to a thousand zebra. These numbers were almost overwhelming. It was fascinating to watch the various animals interact with each other; we saw a quite vicious fight between several individuals, right in the midst of the immense crowd.

We then had a relaxing lunch and afternoon around the lodge, buying a proper leather safari hat for me like the one Jim had bought at Little Kulala. Now we are no longer "tourists." Jim threw me in the plunge pool … memories of earlier years in Hawaii.

The fork-tailed drongo was still in the tree outside our room, and I began to ascribe a spiritual association. I found a perfect area with some especially beautiful, smooth granite boulders near our chalet that had a view out over the woodlands and were shaded by a delicate African star chestnut tree. And I had my own private ceremony in this special spot, to leave some of Kalpana's ash-water in Namibia. I had carried the water to Africa in the same little bag, knit by our daughter Jessica, in which I'd also carried some of this water to India. And now I wanted some of Kalpana to be here, in this fantastic country that we have loved.

We had Amarula on our deck at sunset … a blood red sun, rose-tinged clouds, and Cape turtle doves calling.

At dinner, Sunday brought us a gift. A chameleon that he'd found on his afternoon nature drive. He brought it back to show to us, sitting on a branch, swiveling its eyes, and shaking to pretend to be a leaf in the breeze. Very sweet. Then, after a fantastic dinner we listened to lions roaring during the night. What a glorious – and scary – sound, felt deep in the gut.

In the morning, we saw the lions out by the waterhole, and watched a herd of eland thundering through the trees when the lions came close to them. They raised incredible clouds of dust! We enjoyed watching a few rock dassies on the rocks near the lodge. A couple of babies were hanging on their mother's back but she was rather absent-minded and once shifted her weight so suddenly that one of the babies flew off of her, down into the rocks below!

We were driven to the dirt strip and met our plane, another Cessna 210, which flew us back to Windhoek. We had our last Namibian beer during the three hour layover, and then flew via South African Airways back to Johannesburg for one night. The next morning after breakfast at our hotel, we caught our flight to Lusaka, Zambia.

It felt as if we were starting on a second vacation. It was like a homecoming as we were headed to a country we've been to before and love. We arrived in the airport at Lusaka with great excitement and recognized everything. We also started to look around at the people for candidates to be the fictional characters from the Botswana detective books we love, Ma Ramotswe and Mr. JLB Matekoni. We flew on a charter flight from Lusaka to Mfuwe, and there we met our pilot for the next leg of the journey, and he WAS a perfect Mr. JLB! Sweet, unassuming, looking down at his shoes with shyness, but technically advanced.

It was rather exciting to have a Grand Caravan to ourselves as we were flown to a tiny airstrip in the North Luangwa Park. This park is enormous and very remote. There are only three safari camps located in the park, and each one has only three tents. In fact, while we were there, only our camp had clients, so we were in a park larger than the entire Yosemite national park, with only about 15 humans. There are essentially no roads in the park as it is completely undeveloped, just small dirt tracks leading to each of the camps. We were truly out in the bush!

As we flew, we looked down on the rondavels of the small villages, and when we saw our first baobab tree we grabbed each other with happiness! We just LOVE Zambia!

We landed on the dirt strip where a jeep awaited us. Samson drove us to the camp, about an hour away. As we drove, there were tsetse flies all over the place. We were frustrated at their tenaciousness, just managing to hang on to our skin as we zoomed along the dirt track at 40 km/hr. As we'd never seen them before with the exception of just ONE individual last year, it was a bit of a shock to have to swat at them, and try to keep them from biting. It turns out that this year is a big one for the tsetse fly, and we even had them in the South Luangwa Park later in the week even though there weren't any there last year. I was a bit concerned about sleeping sickness but the camp host assured me that it was "quite unlikely" that we'd get it. So cross your fingers - it has a 10 day latency period before it shows up. When we did complain about them, we were informed that we should THANK them. It's because of the prevalence of the tsetse fly that the Luangwa parks are in existence, since no one could use that land for cattle farming with so many tsetse around! We each suffered about 15 bites, VERY itchy and irritating, but thankfully they stopped itching within the day.

Along the way to the camp we saw impala, Cookson wildebeest, warthogs, and vervet monkeys. And we recognized red-billed hornbills, emerald-spotted doves, white-browed coucal, Lillian's lovebirds, and long-tailed starlings from our visit to Zambia last year.

Soon we arrived at the bank of the river … across from our camp. We had to take off our shoes, roll up our pants, and wade across. Luckily the Mwaleshi River is reasonably shallow at this time of year so there were no hippo or crocodile; even so, it was quite exciting. As the sun had just set, we were invited to join the other guests (Tom and his 22-year old daughter Meg) and the hosts (Rod and Guz) around a fire, and enjoy Amarula for sundowners.

Guz is a Paris-educated cook, and Rod is a white Lusaka native which is pretty rare for Zambia. Kutandala, their camp that they've had for four years, is exceedingly charming. It is a "bush camp," meaning that it's temporary. Every May, at the beginning of the dry season, they build the camp and every November they take it back down. There are four huts, built of sticks and grass with thatch roofs. There is no main lodge, just an area under a leadwood tree with chairs and a table for dinner, along with a hammock and a full library of books on shelves along the branches of the tree. Simple and perfect!

We were taken to our room in the dark, carrying kerosene lanterns, and I fell in LOVE with it. It was one thatch-roofed room with a queen sized bed with mosquito netting hanging from a frame above and a floor of packed earth. The front of the room facing the river was open with only a very low wall - at night a bamboo drape was dropped over the opening to provide protection against marauding animals. The staff opened it back up in the morning when they came to wake us. The back of the room had a set of bamboo shelves for luggage, and a little door in the wall through which they would place a tray of hot coffee or tea in the morning - delicious! There was a doorway to the bathroom which had no roof, a sand floor, and a shower which was actually a 50-gallon drum up on a platform, decorated with dry grass, with a stop cock and shower head. We did have a flush loo. A table with a pitcher and porcelain bowl and another door in the thatch wall, through which they replenished the pitcher with hot water in the morning, completed the bathroom. It was so pleasant to have hot water for washing our faces and contact lenses, and it's amazing how little water a person really needs, after all. There was no sink, just a little pile of stones in the ground where we would pour the water from the bowl. The shower drum was filled with hot water - heated with solar power during the day - while we had sundowners, so we could have hot showers before dinner.

So unbelievably simple and yet so incredibly well designed, and even elegant.

Dinner was the most delicious of the trip yet, I guess the Paris training has something to do with it. But imagine: the closest market is an 8 hour drive away! The "kitchen" of the camp had a full garden where most of the vegetables were grown, and the "stove" was a hole in the ground for coals, covered with a steel plate. There was a second "oven" in the ground for baking bread. Dinner included soup, then stir-fried vegetables (and meat for the other guests) with rice, and then chocolate/orange tart for dessert. Delectable little treats accompanied our sundowners too.

It was quite warm at dinner - finally, we'd gotten far enough north so that we were no longer freezing - and lots of spiders and stick bugs joined us on the dinner table. A much more verdant environment than the desert of Namibia.

We slept perfectly, listening to sounds of myriad crickets and a lone lion calling. In the morning, the "dawn chorus" of birds was raucous. I especially loved the tropical boubou, who made an elegant call that sounded as if someone were running a finger around the top of a crystal glass … and then his mate would reply with a guttural "squawk!"

As there are no roads in the park, the activities of the camp are limited to walking safaris, and we had two on this day, one in the morning right after a great breakfast and returning for lunch, then the second walk was after we had a few hours' of "rest time," and lasted until sundown.

First we went upstream, walking single file behind the scout, who had the rifle, and Rod. Such a varied landscape, deeply wooded areas alternating with extensive grasslands. But these grasslands were different from the ones in Namibia as the grasses were tall, sometimes up to 10 to 12 feet high. We saw lots of puku, the adorable antelope that are only seen in the Luangwa Valley of Zambia, and a few bushbuck who are very shy and thus mostly seen running away. As we walked, we could see Chichendu Mountain in the distance, and even further away was the East Escarpment, a giant mountain range that bisects Zambia. We would stop periodically for Rod to point out aspects of the biology or ecology of the local flora or fauna.

After an hour or so we stopped on the side of a little rise for a refreshing drink of Rose's Lime Juice in water. The scout stepped away and burned some little sticks for heating water in the teapot that he brought along. We then had tea or coffee and a snack of delicious banana bread. We heard vervet monkeys making excited alarm calls, but never found the reason why. We did see some very fresh lion tracks nearby, and lots of elephant and hippo dung.

We ended our walk when we reached the "rhino fence." This is an area of the park which was recently fenced with a 9,000 volt electrical line when Zambia purchased five adult black rhinos to try to repopulate. They had been hunted to extinction in the 1980's, both for the Japanese market, as rhino horns purportedly carry "aphrodisiacal" powers (NOT scientifically true), and for the Yemenese market for dagger handles. SO sad.

The theory is that if they fence the rhinos in for a period of two years, they will each establish a territory within that "safe" part of the national park, which is very carefully scouted with anti-poaching patrols now. This prevents their wandering off and perhaps settling in a territory which isn't as easy to protect. In two years they will take down the fence and hope for the best.

During the walk, we saw 31 new species of birds, many of which we'd seen last year but hadn't yet seen on this trip. Lots of water birds were around us, such as the water dikkop, hadeda and sacred ibis, white-fronted bee-eater, common and wood sandpiper. We loved the piping calls of the white-crowned plover, and the enormous and handsome saddle-billed storks.

Lunch was out under a tree near the river, where there were multiple dishes from which to choose: rocket pesto pasta with homemade pasta, spinach/feta/phyllo casserole, bulghar wheat salad, sweet potatoes with honey and cilantro.

Imagine such a feast in such a primitive place! The dessert was watermelon sorbet served in caramelized candy baskets, all made fresh that day. Wowie.

During rest time Jim slept, but I wrote in my journal and was thrilled to watch 20 or 30 red-billed hornbills hopping about in our "front yard" … pecking and scratching in the grasses. They are such anxious birds, every time I moved slightly they would immediately all fly, flapping and panicking, to a nearby tree.

Around 3:30pm we had tea and apple cake, and then took off down river for our next nature hike. It was delightful. We saw a huge group of baboons on a cliff above the river, which were so agitated by our appearance that they barked at us. What a sound! Then, we walked a long way through tall grass, and could hear nothing but the swishing of the grasses followed by an area of "rain trees" that had lost a lot of their leaves and so we heard nothing but the crunching of huge dry leaves. We came out on the river bank and began trudging through sand. It was JUST like the game we used to play as kids, "going on a lion hunt," where you pretend to be stalking through tall grass, then noisy leaves, and then slippery sand … and this time we really WERE going on a lion hunt! I then had the song "going on a Lion hunt, gonna catch a BIG one," running through my head the rest of the day.

As it got close to sunset, the shadows grew long, and the sun became a brilliant deep orange. A beautiful time of day, very romantic, with puku and bushbuck and impala leaping and Cape turtle doves calling. We came to a glade where there were several crocodiles, and stately ground hornbills strutting about. Then on a nearby hillock, Rod showed us some potshards and part of a millstone he'd found in the dirt, and explained that he'd also unearthed a skeleton at this spot a few years earlier, and it had been dated as 2,000 years old! It had been buried in the Bushman fashion, sitting with the knees drawn up to the chest.

We were so COMFORTABLE hiking along. You might think we'd be on edge, in such a remote place with so few people and so many wild animals all around but to us it felt "at home." I never felt scared, or out of place.

Just then, we had a magical experience. We were walking through a particularly tall stand of swaying grass, enjoying the colors at the end of the day with a salmon-colored sun setting, and suddenly we came out onto the river bank and there, in the water, were two crowned cranes.

I cannot describe the joy I felt. I cried, as Jim held my hand. I had no idea that we would see cranes in Africa. These were the most stunning birds I've ever seen, simply walking along the river, two of them, for us to see. I felt that if the world has beauty like this, then it's really all going to be ok.

Jim said that we were no longer ordinary IMmortals - we are already not ordinary Mortals, since we've seen the albatross.

We met the truck soon after and hopped on for a lift back to camp for Amarula and appetizers, followed by hot showers and dinner by torchlight. This meal we had phyllo filled with lentils and feta; gem squash shells filled with tomatoes, corn, onions, and squash; snow peas; cauliflower in cheese sauce; potatoes. Dessert was toffee cake with custard sauce. Absolutely delicious.

Before bed, I thanked Jim. For our life together where we can see important things like the cranes. And in the morning, we listened to the sounds of guinea fowl and tropical boubou, and I fell in love with Africa yet again.

This day, we took a trip to a different section of the park, at the confluence of the Mwaleshi and Luangwa Rivers, to see hippo. We drove along the dirt track for about an hour, stopping at various waterholes along the way to see birds. It was wonderful to see the fish eagle again, my favorite bird from last year which is now in contention with the bateleur eagle, not to mention the crowned crane. We listened to its haunting and evocative scream. We also loved the massive goliath heron who was sitting next to the tiny green-backed heron, Marabou stork, and hammerkop, and enjoyed listening to the grey louries whining "go-away!"

We saw Cape buffalo, and lots of warthogs - one family even seemed to pose for photos - and zebra. We prefer the zebra here in Zambia, which have just black and white stripes; the more southern zebra wear a brown stripe in between the black and white. We saw many baboons leaping amongst the trees.

As we arrived at the edge of the Luangwa, a large number of crocodiles slid into the water. Such evil-seeming creatures, submerged in the water with only their eyes showing, glaring at us.

And the hippos! There may have been as many as 500 of them, in various pods up and down the two rivers. Most were in the water, some of them moving their huge bulk ponderously about, a constant cacophony of grunting, snorting, yawning, belching and blowing air. They sounded like an old tractor engine trying to be started and we LOVED these sounds!

We then walked down to the edge of the river. I had never thought that we could do such a thing with hippo in the water but Rod explained that in Zambia, the local people have been fishing for centuries alongside the hippo and so the animals are not scared of people. As long as they can submerge, they feel safe. So, we were able to walk right next to the river's edge and enjoy the hippo up quite close. We even walked across the confluence of the two rivers and took photos of ourselves standing in "hippo-infested waters"! It was another magical experience to spend so much time with so many hippos. I took lots of video footage, so we can play it back at home and hear those lovely sounds again.

Our rest time was once more peaceful, with the hornbills hopping and scratching. I did a lot of bird-watching around our chalet and saw orange-breasted bush-shrike, dark-capped yellow warbler, green-backed cameroptera, and black-backed puffback. Most of these I spied while standing on the toilet seat, the best vantage point for the trees overhead! As I was later sitting and writing in my journal, a fork-tailed drongo perched in a tree out in front of our room. I started to write "wow! There's a fork-tailed drongo sitting outside THIS house, too." And AS I WROTE THOSE WORDS, it flew out in front of me and displayed, then went back to his perch. Or should I say "her" perch?

We took another nature hike up along the Mwaleshi after our tea and cake, swishing through the grasses and knocking tsetse flies off of each other with grass stalks grabbed while walking. The person at the end of the line was in the worst shape, with no one to knock off the tsetses, so we took turns in the order of the single file line.

We were accosted by a greater honey guide - another thrilling experience, especially after reading so much about them in my books about the Bushman. These birds want to lead humans to honey, and to do so they will fly around you in circles, calling and enticing, diving and displaying, begging you to follow them. When you do go with the bird, and then reach the source of honey, you are expected to give some to the bird. The most amazing thing about this symbiotic relationship between the bird and humans, is that these are "brood parasites." That is, the eggs are laid in a completely different bird's nest. So the babies are "raised" by parents who know nothing about honey-guiding, and the birds have to: (1) learn how to find honey sources and (2) learn to entice humans to follow them to the honey source, all from instinct. A major mystery of life are these intriguing birds. We didn't want to follow this bird to the honey, but he didn't give up easily. He kept after us for at least 20 minutes, incessantly calling and begging which was really surprising to us.

Climbing a hill at sunset, in the distance we caught sight of elephants, a mother and two babies, coming down to the river for a drink. Magical …

Back at camp, Guz had set chairs out in the river for our Amarula sundowners, and we watched bats fly overhead and listened to the water dikkops with their slowly decreasing plaintive song - nicknamed the "dead battery" bird. Dinner was incredible again: lettuce and lovage soup; sliced cooked turnip; mashed sweet potato with eggplant/tomato/olive sauce; sweet potato chips; individual lemon soufflés and homemade guava ice cream served in a large bowl made of ice. All out under the huge sky of stars, the Southern Cross overhead.

I asked for an early wake up, 5am, so that I could see Orion and Venus before it got too light. However, when the fellow came to call "good morning!" I was too scared to go outside of the chalet in the dark as I could hear hyena quite near. I was glad that he also opened up the front blind so that I could go "outside" and yet still be within the little wall out front of our room. And I was able to see my beloved constellation.

This day after breakfast we drove for an hour along the dirt track in the opposite direction, towards Chitenge Mountain. There was a spider spinning a web from the radio antenna of the jeep towards the windshield, which stayed intact even as we were zooming along. "Extreme web-building" was Jim's description of the new sport.

We stopped the jeep under a large tree, and began our hike to the Mwaleshi waterfalls. This time we not only had Rod and the scout, but we also had a "porter" along with us which made me feel as if I were a character in "Out of Africa" ... an intrepid woman explorer, with her husband and the carriers following along.

I wished for a fork-tailed drongo, and saw one. Hmmm. Makes one REALLY begin to wonder...

Jim and I had a lot of fun while walking, turning to look at each other periodically, and teasing each other a bit. He gave me a pretty grass stem to put in my hat band, and then called me "Puss! – in boots," from the movie "Shrek 2." Another time he was goosing me and I reprimanded him with a downpointed finger, from the romantic movie "Dances with Wolves." Precious shared moments.

We saw a line of safari ants and collected some interesting seed pods to bring back home. We saw several new kinds of birds, including one of the most ancient birds, a purple-crested turaco, which hops to the top of a tree and then glides down, just the way prehistoric birds began to fly.

We suddenly came across a HUGE baobab tree. Gorgeous and primeval, it seemed SO alive and intelligent, even kind. We found one of its fruit, and sucked on its bitter lemony taste. It was dry, like astronaut ice cream.

Next in our path was a sacred sausage tree that apparently had the bones of a tribal chief buried in it, from several hundred years ago. All the elders of the local villages travel to this tree every October to bring white strands of cloth to tie in the branches, and to ask for the chief's blessings. They say that when you pass the tree you must pay homage so that the chief would protect you from harm. We knelt next to the tree and each in turn threw a small amount of mealie meal in a cleft in the trunk, and exclaimed, "Thank you, Chief Mukungule!" The ceremony was reverential and special. As we walked away, a martial eagle flew overhead, the Chief's acknowledgement according to the scout.

We had our tea and cookies on a cliff overlooking the river, with a rather large group of hippo standing on the opposite sandy riverbank. Rod and the two guys tried to frighten them into the water, and it took a lot of effort ... gesticulating, throwing stones, shouting, even climbing down the embankment … finally the hippos did decide to "rush" into the water. It's amazing how fast a hippo can actually move!

We then hiked up to the top of the falls, and had a lovely time relaxing and eating lunch by the river. The porter had even carried cold beers for our pleasure! They built a fire and cooked kabobs, and I had a special foil packet of veggies and cheese, with several salads, complete with tablecloth spread on the sand. We went scrambling over the rocks, and found that the river makes multiple pools as it cascades over several different sections of rock. Rather like the "seven sacred pools" in Maui. Rod explained that we could swim in one of the pools, and this one was a whirlpool. It was really fun to jump in and go around with the whirl, but we had to be really careful not to slide over into the next lower pool. Not only were the rocks sharp, but there were sometimes hippo in that pool, as Rod found out the hard way once, and was lucky to survive. Jim did slide into the whirlpool, which was quite brave, in my opinion, but he scraped his hip on a rock so we had to bandage him up a bit.

After a few hours of playing and relaxing we walked the few miles back to the jeep. On that walk, Jim and I were being silly once more, and at one point while I was looking back at him and making faces, the line stopped in front of me to look at a bird and I didn't notice until I smashed into Tom, and then he smashed into Meg, and the whole line became like dominoes. I was caught!

On the drive home, we saw a slender mongoose and then a family of banded mongoose (mongeese? mongi?), one of which stood up in the road ahead, like a meerkat. At sunset we came across a group of elephant right near the road and it was great to watch them "circle the wagons" as they formed a ring, facing outwards, with a tiny baby inside of the adults. The females challenged us with their ears flapping for a while, and then they all turned and trundled away. Magical, again!

This night, after our Amarula sundowners and hot showers, we had dinner out on the sandy river bank, surrounded by torchlight. Very romantic … and it was a delicious Indian meal. Lions called in the distance.

Before dinner, Rod came to tell us that he'd gotten an email saying that our son Mark wanted us to call but that it "wasn't an emergency." The message was already a few days old; Rod and Guz only check email by satellite a few times a week. So I asked to use their satellite phone; Rod said it was literally never used but he'd let me try. It was patched through Paris and cost $5 per minute. So while Jim and I speculated about what was wrong with Mark (did he run out of money? burn down his apartment? get arrested?), I dialed the phone. And Mark answered, and explained that he'd simply had an unsettled feeling that something was wrong, and wanted to make sure that we were ok. Wow. I guess he actually MISSED us! We were VERY touched.

The next morning we had a quick breakfast, and then a drive out to the landing strip, catching sight of a gorgeous spotted hyena on the way. Our plane landed at exactly the same time as we drove up to the strip, and we sadly said goodbye to Rod … we had loved our time in the North Luangwa.

We flew south to Mfuwe, at about 500 feet over the ground, and enjoyed the views of the hippo and buffalo, and the baobabs. At Mfuwe airport, we were met by the truck from Kaingo. On the drive through the town of Mfuwe, which we love, so many industrious people strolled along the main road, carrying buckets of water on their heads, riding bicycles, chatting and laughing and waving. We asked to stop at the Textile Factory and although it was Sunday, we got a little tour of the rooms and watched how they decorated, painted, and created the beautiful batik fabrics, then bought some wall hangings, place mats and other gifts in their gift shop.

As we entered the South Luangwa Park we were filled with excitement. We'd been here before! To greet us, there was actually a fork-tailed drongo at the entrance gate. We saw zebra, impala, puku, elephant, giraffe, waterbuck, and lots of birds. Fantastic.

We were driven to the bush camp of Derek Shenton, called Mwamba. Derek is also the owner of Kaingo, the safari camp that we loved the most last year, and so we booked one night at the bush camp and then our last two nights at Kaingo itself. Derek is quite a character; another one of the few white Zambians and whose farmer father is also from Zambia, his mother is from Switzerland. His parents were at the camp when we were there last year, and it had been fun for me to meet them since I'd actually read about them in the book, "Don't Let's Go to the Dogs Tonight."

Mwamba has three nice little chalets, all made of sticks and grass. There were red-eyed doves and Cape turtle doves calling, along with a whole host of other birds, so we were quite content.

After a nice lunch, we went for an afternoon game drive with our guide, Patrick whom we remembered from last year. Patrick is a native Zambian who lives in a rondavel home in Mfuwe. There were also two other guests, an older couple from England who were quite nice and seemed full of fun, Roderick, the fellow who would hold the spotlight later in the evening, and Emma, the young camp hostess came along for the ride. We were in an open-topped jeep as always, Roderick and Patrick in front, the other couple in the second bench seat, with Jim, Emma and I seated on the back bench.

It was a nice drive, slow and careful, enjoying the scents and sounds of Africa along with the sights, savoring every aspect of the wilderness. We saw three male kudu crossing the road and sat quietly to watch them slowly move by. We stopped for Amarula sundowners at a small pond covered with lilies, and watched jacanas picking their way amongst the lily pads, seeming to walk across the surface of the water.

After sunset we started driving again, the older coupled teased Patrick about the fact that they'd been in Africa for three weeks and hadn't seen a single lion … they REALLY wanted to see lions!

We drove and drove and drove, and Roderick scanned with the spotlight from left to right, then back again, and back again … it was like watching a tennis match … and all we saw were four porcupine, which were actually quite adorable. We continued on, still in search of lions.

And then we saw a lioness! She was stalking some buffalo and we watched her for a short while. She even made a half-hearted attempt at running after them, up a little rise, but quickly gave up. She needed some other lions to help with the hunt, and as she walked along she began so softly to call to her other friends. We drove up right alongside her and stopped so Jim could get some good photos.

After a few minutes, we wanted to drive further forward, to get a better photo from the front, so Roderick turned off the spotlight, while Patrick turned off the headlights and turned the ignition switch.

Nothing.

He turned it again.

Nothing.

At first we thought he was joking, but then we realized that the battery was not sending any juice to the starter. Or the headlights. Or the spotlight. Or the radio.

And that's when I began to get scared. Along with everyone else. Patrick exclaimed in a nervous voice that he could barely see, but that some other lions had joined the first one, and were walking around the jeep. He was especially concerned that some were even BEHIND the jeep. Right behind US.

We were all nervous and asked whether he had a flashlight, or a hand-held radio, or a GUN. He told us he had his panga in his hand. Oh, great, a machete against lions. Jim held our camera mounting bracket as a weapon too, thinking this would comfort me. Right.

I got so frightened that I sat down on the floorboards of the jeep, but when I thought about the fact that then Jim might be eaten first, I got back up on the seat.

We all just sat there, stunned and worried, for about 10 minutes. We weren't expected back in camp for hours, and even when we didn't arrive it would have been hard for them to find us, as this was a huge national park and no one knew exactly where we'd be.

All alone. In the dark. Surrounded by lions.

Jim had been thinking instead of just worrying like the rest of us, and suggested that we use the 12-volt battery from his camera to power the spotlight, so we could see if the lions were still there. So they cut the wires to the spotlight and handed them back to him. He got out his reading glasses, and asked everyone to turn on their digital cameras and shine the displays towards his hands, so he could distinguish the + and – terminals of the battery. His plan WORKED and we finally had light ... and discovered that the lions were indeed STILL THERE. And, there were five of them. Two on our left side, about 10 yards away, and three in front, about 20 yards away. It was nice to see that they were all lying down, but they were still all staring at us.

Then Jim suggested that the guys shine the light up in the sky, "SOS," to try to get the attention of any other jeeps that may be out there. They did shine the light up and wiggle it about for a while - not entirely clear on the "SOS" concept - but it was pretty obvious that there were no other vehicles. It was completely pitch black.

Patrick assumed that the battery terminal had come loose, but there was no way to get up and under the bonnet, with the lions all around. They don't mind jeeps filled with people, but if you get OUT, they mind. And they will attack. They are opportunist hunters and will kill if food presents itself.

Jim suggested that he use the camera battery to power the radio. The guys up front weren't sure how to get to the wires, so Jim climbed up and over the seats to help. I suggested that Roderick may want to come back and sit by me, "since it was so crowded up there now." Of course it had NOTHING to do with the fact that the seat next to me, nearest the lions, was now empty. But Roderick was too scared to move. A few minutes later, Emma politely asked if I could get off her LAP! Guess I was migrating closer and closer to her - that was a bit embarrassing.

Anyway, Jim's plan worked (I was SO proud of him), and Patrick called for help. Actually, he calmly told Derek that we needed a bit of assistance with our battery. We were all begging him to scream "HELP! HELP! We're surrounded by LIONS!" but he needed to maintain his professional demeanor. He had his pride to consider (interesting choice of words/no pun intended).

While we waited, we decided to raid the liquor stores – we felt we needed a little "liquid courage" – and managed to completely finish the Amarula and gin in the cooler in order to brace ourselves.

Derek eventually came out with another jeep some 40 minutes later. With all the lights and the commotion the lions quietly stood up and wandered off. Oh, sure, NOW they leave.

It did turn out to be trouble with the battery terminal, and Derek and Patrick fixed it. Driving back to Mwamba, Patrick kept finding interesting animals - elephant shrew, banded mongoose, hyena - and we all kept saying, "yeah, yeah, fine, don't turn off the engine, heh heh."

When we got to camp, the staff came out to explain that there were elephant in camp. At first we thought they had to be JOKING, but there actually were two elephants munching the bushes right by the huts. So we stayed in the center of camp, having drinks at the bar, laughing nervously and beginning to relax.

"SO, you wanted to see LIONS, did you?!" we chided.

After a lovely dinner out under the trees, we were able to go to our rooms as the elephants had wandered off by now, and we had a great sleep. Lots of insects sounded all night - at dinner I'd asked "What bird is that calling?" and Patrick had answered, "A cricket."

We were woken at 5:30am, had tea and biscuits, and were off for a game drive. This time it was just Jim and me with Patrick, and it was great fun. We watched a pied kingfisher with a squeaker fish in its bill, repeatedly smacking the fish against a tree limb to break the fish's spines off. Egyptian geese parents with 6 babies and a family of banded mongoose with lots of babies running through the bushes delighted us, and we even watched an impala leap OVER another impala! We came across one of our lion friends from last night, trotting along and one was carrying a LEG in her mouth. It appeared to be impala or puku; Patrick was grateful that it wasn't HIS leg. It was horrible to listen to her munching and crunching it.

I asked for a baboon sighting, and was delivered a wonderful one. A huge troop were gathered in a tree by a small waterway and as we pulled up in the vehicle they started to jump down from the tree. One by one they leapt over the watercourse. Even the tiny babies were LEAPing in the air! Some of the adults had babies clinging to their tummies or backs. It was sensational.

We drove on to Kaingo camp, where I cried to be back again, and devoured a delicious full breakfast. After that we went to the elephant hide down by the river and enjoyed watching an "argument" between two matriarchs. There were a group of about 12 elephant drinking in the river including some adorable babies. After they finished, one of the elephants started towards our side of the river, while another one went the other way. It took a long time for each of the other elephants to "choose sides." Some went one way and some went the other. For a while, it appeared that "our" elephant was going to "win" as 6 of the 10 were coming our way, but then something happened and the ones that had been moving towards us slowly turned back the other way. We could hear some of the communication grumbles, but most of it must have been in the sub-sonic region. Soon, only the one elephant was turned towards us, with her nursing baby. Eventually she too gave in, and turned and followed the others. It was really interesting to watch the contest of wills!

After a yummy lunch and a bit of a nap, we went on our afternoon game drive. Again Patrick was our driver, and we noticed that this time he brought a hand-held radio and a flashlight! We saw vervet monkeys, and a trio of zebra that seemed to vanish into the thick green grass. We startled a baboon mother and her child, who were on opposite sides of the road, and so they started to run towards each other. The mother ran across the road in front of us and grabbed the baby up in her arms, mid-stride, all the while shrieking. They are apparently devoted mothers. While having our sundowners on the side of the river, we saw two crowned cranes flying overhead.

On the night drive, we saw 6 elephant shrew, five genets, and three porcupines. Then, an evil crocodile crossing the road, quite eerie looking, out of the water, and also a beautiful civet. We saw the five lionesses from last night again, sitting down, relaxing on the grass and were able to take photos of them this time.

Two gorgeous leopards came into sight, which was fantastic! The female was pursuing the male - she was in estrus - and we were able to watch them walking along together for quite some time. Magnificent animals.

After a scrumptious dinner we had a marvelous sleep, with the sounds of hippo all night long. Yeah! And I was able to see Orion and Venus in the early morning, out the front window overlooking the river.

I must admit, however, that I was really beginning to miss the kids terribly and we decided that next year on our Africa Trip we'd arrange for me to call them a few times so that I don't get so sad.

On our last day of safari, we had our early morning game drive and enjoyed looking for leopard. We kept hearing monkey alarm calls as we bumped over the uneven surface of the ground caused by hippo tracks. We saw many herds of zebra out on the plain, and got fantastic photos of lilac-breasted rollers ROLLING and spinning and looping. Two males competed with each other for the attention of a female, and were performing incredible acrobatics. For our coffee break, we watched a pond filled with crocodile who were causing the fish to leap out of the water. There was a clever group of yellow-billed storks waiting in front of the fish, grabbing them as they leapt. We saw a zebra taking a dust bath, and three bateleur eagles soaring overhead - a juvenile, female, and male.

Before lunch we sat down by the river and saw a 4-way baby hippo play fight! They were SO adorable, as their little snouts would come poking out of the water, biting at each other and rolling around and around. We ventured to the hippo hide up the river and watched the adult hippos submerged in the river for an hour or so. It was quite remarkable to watch their dynamics as one would shift in position every now and then and this disturbance of the status quo would either make another hippo angry, in which case there was a bellowing confrontation and a pushing war, or a whole group of them would stand up and loudly rearrange themselves, with grunting and rumbling and snorting. It was also interesting to see the brilliantly colored oxpecker birds hard at work on the hippo's backs, pecking at insects or the blood in their cuts - that part was gross to watch.

On the way back to camp, we passed right next to an elephant who was wandering around near our chalets and chomping the bushes and grasses. We watched her grab a bunch of grass by twining her trunk around it, but since it still didn't come free she also used her FOOT to shear it from the ground. Quite skillful!

After lunch and an enjoyable afternoon rest listening to hippos, we started out on our afternoon game drive. This time Derek drove us, and the first bird we saw was a fork-tailed drongo. So I knew it was the right time to have my Zambian ceremony with my Kalpana ash-water. When we stopped at a special location on the riverbank near the ebony forest, I poured the rest of what I had carried to Africa on the bank of the Luangwa River. When the rains come, it will be washed out to the Indian Ocean, along with the rock that I brought from our yard and placed alongside. I miss my friend.

We saw a leopard cub walking around through the bushes, probably out on one of his first solo trips, with monkeys crying in alarm overhead. A giraffe couple slowly wandered along the road, eating from the tallest trees. As it got close to sunset, it didn't appear that Derek was going to stop for sundowners, as he simply seemed to drive around faster and faster, and Jim got more and more worried about missing the sunset. At one point, the sun even dipped below the trees, and Jim and I exchanged anxious glances. Earlier I had asked Derek to stop if we saw some impala leaping so that I could film it, and when Derek pulled over near some impala at one point, we practically screamed at him to forget it, that the sunset was more important! We came around some bushes … and there was a table set up on the bank of the river, complete with white tablecloth, candles, Amarula and snacks! And the sun was just setting. We got video footage and photos of our last sunset, holding our drinks and smiling (and crying).

The new moon rose, like a perfect bowl in the sky – or a smile.

On the drive back to camp in the dark, we enjoyed a Pel's fishing owl, a rare sighting, with feathers blowing in the breeze, and then caught sight of our little leopard cub again - just a fleeting glance before we lost him in some bushes. A few moments later, we saw a hyena, and then we saw the hyena CHASING the leopard! It was very exciting and we were all relieved a few minutes later when we found the hyena walking along, ALONE. Luckily he'd not gotten the little cub; that would NOT have been ok with me! We saw a civet sitting in the crux of a tree branch, looking for all the world like a little kitty, and several more hyenas running in their guilty-looking limping style.

Our "final supper" was outside on the bank of the river with a delectable barbeque. We slept soundly this last night to the sounds of hippo and hyena.

Our early morning start had us heading for the airport at 5:30am for our (tortuous) trip home. On the way, we saw a huge gymnogene - the largest eagle - hanging upside down from a tree branch and then hopping up along the trunk, peering inside squirrel's nests for a snack. And a fuzzy baby zebra, one month old, suckling. And a last fork-tailed drongo.

We drove back through the sweet town of Mfuwe, passing mothers with their babies in slings, and smiling children with pails of water on their heads, waving and calling

"GOOD BYE!!"

Chapter 3
Africa 2005

✦

We're back from another FANTASTIC trip to Africa!

This time we decided to have a different kind of trip. Instead of traveling from country to country and seeing many different ecosystems as we did in our previous trips, we focused on a single region of a single country. This was partly for economy and also because we realized we had been spending a lot of time in transit.

It was hard to decide which region to focus on. We had a serious conversation about what we loved most in our previous trips and although the first answer was "everything!," we tried to be more selective. We came up with:

1. Watching elephants
2. Listening to hippos

We've read several books about elephants (or, "elies" as many people like to refer to them) this past year, including," The Eye of the Elephant," and "Elephant Memories," and are completely in love with them, particularly because of their fascinating and complex social structure:

- Each herd is led by a matriarch - I like that!
- Infants are so close to their mothers and aunts that they are never more than a few FEET away from them for the first 4 years of their lives.

- Female elies stay in their family groups for life, which can last more than 60 years.
- Each group of females and youngsters, referred to as a "breeding herd," are extremely intimate with each other, squealing with joy and delicately touching each other's faces after short separations while foraging. They also raise the young communally, as sisters step in to help young mothers with child care.
- Young bull elephants leave the group as teenagers to go hang out with "bachelor herds" of guys.
- Often, a young bull elie will spend time with a particular older bull who will teach him the skills he needs, and they'll develop a special bond. When the older one becomes aged, the younger one sticks by and takes care of him.
- Elephants are very respectful of each other and exhibit sadness when a family member dies. They will carefully "bury" a slain elie with tree fronds, and when they find elephant bones they smell them, caress them, and return to them again and again, but not bones from other species.

We wanted to spend time watching these behaviors, now that we'd learned so much about them. Their ability to communicate sub-sonically over great distances has also been recently discovered, and we knew that we'd probably be able to hear SOME of their rumbling and realize that communication was occurring.

But further, I also had a personal reason for wanting to spend more time with the elephants. I had always thought of them as one of the greatest of the world's creatures, and was completely drawn in by the babies in particular. But during our first two trips to Africa, I had suffered quite some fear in their presence. Several times we came upon a group of elephants unexpectedly, as when driving through rather thick brush and their reaction would be to abruptly face us, fling their immense ears straight out to the sides, and trumpet wildly.

This was scary to me!

But in our readings this year, we learned that in fact this behavior is no problem at all. They are simply trying to scare you. They have no intention of causing harm, they are just offering a warning that you should come no closer.

It's when an elephant keeps its ears tucked in and trunk tucked under (for aerodynamic reasons), and comes straight at you, that you have to worry. This means war. She plans to kill.

Armed with this knowledge, I was looking forward to getting trumpeted at, and NOT being scared! We've simply got too much video footage from our previous trips where I'm looking really anxious. This time, I planned to revel in their bluster with pleasure.

We researched extensively to find which parts of Africa had the highest concentrations of elephants, and were happy to read that the Lower Zambezi National Park in Zambia was a good choice. We'd been there before on our first trip to Africa and had stayed four days at a camp called "Sausage Tree Camp." We are really fond of Zambia. Furthermore, our other wish, to hear hippos, would also be satisfied in this national park as all of the camps are directly situated on the Zambezi River (of Stanley and Livingston fame), along which there are approximately 80 hippos per kilometer. Yeah! Gruntings and burblings all night long.

We worked with our wonderful travel agent, Bushtracks, to set up a trip with 14 nights in the Lower Zambezi National Park, at four different camps along the river. As always, they arranged everything for us, really, everything! All we had to do was show up in Lusaka on August 6, 2005.

We'd spent weeks preparing for the trip. We already had all our safari clothes and hats and even our necklaces from Namibia, but we wanted to make sure we had the most useful arrangement of cameras and video, and all the required batteries and video tapes and flash cards, etc. It sounds as if I know what I'm talking about; however that was all Jim's department. I focused on setting up our appointments at the clinic to get our shots and malaria pills, and making sure we had our airline tickets and passports. We also spent a lot of time organizing our toiletries to a bare minimum. We wanted to have minimal luggage so that we wouldn't have to check anything.

Jim had also built something special and new for the trip. Last year we noticed how we wanted to look at the photos we'd taken during the day so he built a briefcase that contained a computer and photo printer, powered by solar batteries. Incredible! This was to turn into quite the item on our trip, as will be seen later.

Our actual trip started on Thursday, August 4, at 4:30am, when our friend Diana came to drive us to the airport. This year we were very lucky to have accumulated lots of airline miles so we opted to use them to travel business class for free on American and British Air. It was wonderful. The seats turned into beds, and we were actually able to sleep. We are now completely spoiled. Just my WRITING about it is making me get excited all over again!

With an entire day layover in London we were able as usual to have lunch at our favorite restaurant, Rasa W1. This time we'd called our friends Chicco and Ludovicka, whom we met in Namibia last year, and invited them to join us for lunch. Chicco had a new job so he couldn't join us but it was nice to meet up with Ludovicka. After a delicious meal, we decided to go visit 221b Baker Street. For anyone who knows the significance of that, I had recently been reading out loud to Jim, the entire collection of Sherlock Holmes stories. That was fun! Then, it was back to the airport, relieved that we were done with our traveling via subway as it had been bombed only two weeks previously

Our flight to Lusaka was uneventful, as we slept blissfully and arrived at 6am. As we stumbled off the plane, bleary-eyed, there was a guy holding our sign:

TUTHILL X 2

Made me cry!

David Cumings met us while we were in line for customs and introduced himself as the owner of our first camp, Chiawa. He spent quite some time chatting with us, telling us the history of the camp. It was the first camp in the National Park; in fact he and his wife were involved in setting up the National Park through their friendship with the former President of Zambia and it's been in operation for about 15 years. His son Grant now takes care of running the camp, while Dave lives in Lusaka and gets to welcome arriving guests. I thought that was a very nice touch. He helped us up to the terrace of the restaurant before leaving, and ordered coffee/tea and sandwiches for us. We had about an hour and a half before our next flight, and it was quite charming to sit on the terrace. We even saw our first birds: house sparrows! How exotic, we joked.

We boarded a 12-seater Cessna Caravan and before too long, were again landing at Royal Airstrip, a dirt strip in the middle of the bush, and once again, there was a jeep waiting to pick us up.

We felt as if we had arrived home.

When we landed on the dirt strip and pulled over to a nearby tree to unload, we saw a small hangar next to the runway with an awesome plane in it - a Bellanca Scout. Jim and I both blurted out, "We can FLY that plane!" Peter, the driver of our jeep, explained that the plane is owned by "Conservation Lower Zambezi" (CLZ, pronounced by the locals as "seal-zed" due to the British pronunciation of the letter "z" as "zed" rather than "zee"), and is used for surveying as well as finding animals that need help. More on that later.

Royal Airstrip is situated outside the western edge of the Lower Zambezi National Park, which is aligned along the east/west running Zambezi River. The Park, which is downstream and east of Victoria Falls, extends along the river for 70 miles; the river itself is the border of Zambia and Zimbabwe. North of, and parallel to the river, is a beautiful rugged escarpment; this acts as a physical barrier to most of the animal species, which are therefore concentrated along the valley floor.

A 15 minute drive took us to the CLZ buildings, where I got to hold an orphaned baboon who was only a few weeks old. CLZ was transferring him to an orphan-rehabilitation group that has great success re-introducing young animals to family groups in the wild. We hopped in a boat, traveled downriver and further east for another 15 minutes before pulling up to the boat dock at Chiawa Camp. We were welcomed by the two camp managers, Barbara and Craig, and a smiling staff member holding a tray with ice-cold towels for our hands and faces.

It must have been entertaining to watch our eyes grow big as they gave us a tour of the main lobby area more commonly referred to as the "sitenge," with a bar and lots of comfy sofas and chairs, both inside under the thatched roof, and outside on the grass, as well as upstairs on an outside deck. The building has only one wall, back towards the bush; the other three sides are open and look out across a sloping lawn down to the river.

After a refreshing drink and an incredibly delicious lunch, which tasted even better with our first taste of Mosi, a great Zambian beer, we were shown to our chalet. Barbara proudly announced that we were in the honeymoon tent, and Jim and I melted. It was perfect. Sumptuous, yet simple, a large canvas safari tent set on a raised wooden platform was presented to us. Inside we found a most welcome queen sized bed surrounded by mosquito netting and bedside tables, as well as a table and chair. The front and back of the tent had zippered netting, likewise the windows along the sides. The deck out front sported chairs and a table. The bathroom was out back, and enclosed by wooden walls about 7 feet high, but with no roof. Yeah! Birding in the bathroom again! A great outdoor shower and loo (listen to me, already comfortably using British words like "loo"!), double sinks, and being the honeymoon tent, a gorgeous claw foot tub completed the suite. A window looked out over the dry riverbed of the Chiawa River tributary and leading down to the Zambezi.

We stayed at Chiawa for four days and loved every minute.

Each day started at 6am with a "Good morning!" as a tray of coffee and tea was placed on our deck. Half an hour later, we gathered around the fire in front of the sitenge, watching the brilliant red sunrise over the river as we ate breakfast. This usually included bread toasted over the fire, granola, fruit, and porridge - flavored with Amarula or rum no less! By 7am, we left for our morning activity (we had a choice of a driving safari, a walking safari, a river boat safari, or canoeing), and returned to camp around 11:30am for brunch. Of course, we also had been given coffee/tea and cake halfway through the morning so that we wouldn't pass out from starvation.

Brunch was an incredible affair with multiple offerings: omelets made to order on a hibachi, home fried potatoes, warm freshly baked breads and muffins, curried vegetables, some with meat, fresh salads, sliced fruits, and cheese plate. No wonder we both gained weight on this trip.

We then luxuriated in our afternoon siesta until 3:30pm, at which time we gathered again around the fire and had yet another freshly baked delicacy with more coffee or tea, and then left for our afternoon activity. The most important question at this gathering was what we wanted to have for sundowners. For us, the answer was always the same - Amarula, of course.

During the afternoon activity we always stopped at about 6pm to have our drinks and snacks, and ceremonially watch the brilliant red sunset. Sundowners were followed by a night drive in a jeep, searching for leopards and other nocturnal animals, until around 8pm when we arrived back in camp for drinks around the fire. After excitedly discussing our day with the other guests and guides, we were serenaded by the "Chiawa Choir," a group of about 12 staff members who marched down to the fire and sang enchanting African songs sounding like the music on Paul Simon's Graceland album, ending with a song inviting us to dinner.

As was fitting, the table was set outside under the stars, lit by glowing candles and kerosene lanterns. Dinner was announced by the chef: soup, main course, and dessert. I was well catered to with my vegetarian preference and always had a separate vegetarian main course that included multiple vegetables and rice. Wine was liberally served throughout. The dishes were tasty and complex, made from the freshest of ingredients grown in the local villages.

After dinner, we were escorted back to our chalets by a staff member who was more knowledgeable in dealing with the various elephants, hippos, buffalos, or lions that may be wandering about. Our tent was magical – the bed was turned down, and inside, two hot water bottles. Each room was lit with kerosene lanterns. We fell into bed for a wonderfully peaceful sleep and a night filled with hippo sounds - yeah!

Back to our first day at Chiawa, we arrived at siesta time. We thought we ought to sleep, after the rather tiring trip and the jet lag. We enjoyed our first, fantastic, outdoor showers, dressed in the bathrobes provided in the tent and laid down on the bed. But we couldn't STAY there! As we looked out across the dry riverbed we saw a group of warthogs arrive, who began to forage in the grass down on their front knees. Then, a male and female bushbuck wandered out from the trees at the other side of the riverbed, gracefully bending down to graze. A large group of impala moved into view, some leaping and pronking (springing straight up from the ground). And when the baboons arrived, a whole troop of about 50 individuals, running and leaping and bantering, carrying babies on their tummies or backs, I realized I simply could NOT spend my time sleeping! I had to get up and watch. We sat out on our porch, binoculars and cameras in hand, and watched the incredible pageant unfold.

Our other siestas were similarly - and delightfully - interrupted by interesting animals. Near our tent was a blind built in a tree where we spent several hours watching elephants taking mud baths right below, and hippos wading out of the river and onto the grass to munch. There were also delightful birds to see: white-crowned plovers and lilac-breasted rollers and fork-tailed drongos.

One afternoon was particularly exciting. Jim had gone over to the sitenge to print some photos and to find a good spot for the solar panels. That was proving a challenge since our tent was surrounded by beautiful mahogany trees and the small amount of direct sunlight kept shifting location. I took a lovely hot shower, intending to wander over to join Jim. As I entered the tent from the bathroom, I heard a very loud rustling in the trees to my right. I walked to the front of the tent and peered out through the screen.

THERE was an elephant!

He was walking towards our tent, from right to left, about 2 feet away from the front of our deck. I held my breath and watched as he sedately walked by, swinging his trunk, and glancing from side to side. When he was directly in front of our tent he turned his head and looked RIGHT AT ME. I was shivering with excitement, and kept repeating to myself, "They don't come inside structures, they don't come inside structures ..." He stared at me for a full minute. Then, he simply turned his head back, and kept along his way.

It was a true encounter. I was speechless and thrilled.

I couldn't wait to go tell Jim. I grabbed my binoculars and hat, unzipped the front fly, stepped out on the deck, and turned around to zip the fly back shut. It was just then that I heard some more loud rustling in those same trees. I glanced over, and ... YES! ... there was another elephant coming!

I calmly, well, probably not calmly, unzipped the tent and stepped back in. This time I was looking out between the open flaps, with an even better view of this elie as he too, stopped, turned and stared at me. While this encounter was going on, and before he turned to continue his walk, I heard some sound on the path to the left of the tent. I glanced over, through the window on the side of the tent, and there ... striding along with his head down ... was Jim.

Heading straight for the corner of the deck, where he would meet face to face – or rather, face to tusk - with a large bull elephant.

This would not be good.

I called out quietly for him to STOP, then I called a little louder, and still a little louder. Luckily, he heard me before getting to the front of the tent ... and stopped. And I had saved my husband's life! He said that he had been very carefully scanning for animals as he walked along the path, but then relaxed and dropped his guard when he arrived at the back end of our tent.

A lesson to be learned: never drop your guard in the bush.

These bull elephants seemed to visit the camp nearly every afternoon and we happily came across them many times, never forgetting to keep our eyes peeled. It was marvelous to sit in the main room sipping on a Mosi, watching an elephant munching on a nearby tree.

One lunchtime we were directed down to the dock and there, loaded on board the boat, was a complete luncheon event: a waiter, a cook, and even a hibachi for making omelets! Barbara popped open a bottle of champagne and we toasted to Jim's birthday. So sweet! We drifted languidly along the river, eating, chatting and taking pictures, before returning to camp for siesta.

We went on several game drives during our time at Chiawa. Each one had exciting and special moments. We were usually the only people in the jeep with the guide and the spotter so we were able to stop often and do a lot of intense birding. We saw 81 species of birds, 17 of which were life birds. The guides were remarkably knowledgeable and very respectful of the African bush, which we appreciated.

A good deal of time we spent watching elephants and got to experience our first trumpet: I wasn't scared! As is well published by now, I especially loved to gaze at the babies. One infant that we came across was only a week old and still wobbly on its legs which was so entertaining while very endearing. We truly also enjoyed watching the other families: baboons, monkeys and warthogs, and seeing bushbuck, waterbuck, kudu and impala.

One day while out for a lion hunt we passed another jeep that informed us that they had seen lions over near "the red cliffs." An exciting drive ensued, the red cliffs were quite pretty, but we didn't find any lions. We DID track their spoor some distance, leaning out of the jeep and staring at the prints in the dust until they disappeared into some dense thickets.

Our first successful sighting of lion was during a night safari. This encounter also involved a tremendously long search, after we were given a tip that there were lions over near Sausage Tree Camp. The guide said there was one sleeping IN a sausage tree, which is considered rather unusual for a lion. We drove to the area he mentioned, and exhaustively searched in every single sausage tree ever known to mankind. Or at least that's what it seemed like. We finally gave up and enjoyed the night sky just FILLED with brilliant stars; the Milky Way, the Southern Cross, Venus and Jupiter, and a tiny orange crescent moon.

As luck would have it, just as we were driving back to camp, we essentially bumped into a group of lionesses - six of them striding along together in a group. They appeared to be hunting so we sat quietly with the lights out waiting to see if they'd be successful. We heard lots of impala giving alarm calls (a high pitched bark), but no chases. We also shone the light on them for a while and enjoyed their camaraderie as they nudged, and nestled, and jumped on each other, and swiped at one another with their huge paws. It was entertaining to watch them play. Lionesses are beautiful creatures, with soft beige fur, dappled at bit towards the back legs.

That night we had a romantic bubble bath, surrounded by candles, listening to the hippos … while a buffalo munched right outside the window. I did say, romantic, didn't I…?

On our night safaris we saw lots of fascinating smaller animals: porcupines with their tiny little legs and adorable little faces, hyenas, and genets which look like tiny leopards but are actually members of the mongoose family. They have spotted bodies with long beautiful fluffy tails. One night, we first heard and then found a male lion snoozing. He'd been given the name "Douglas" by the guides, and although we KNEW he was the king of the forest, he didn't look too scary as he lay on his back, snoring, with legs splayed. After our night drives we were always welcomed back in camp with warm towels, then ushered to the blazing fire to enjoy a drink and share stories of what we'd seen.

One morning we opted for a walking safari. As we remembered from our past trips, it's all about DUNG and TERMITES! Nonetheless, it was fascinating. Our guide, Joe, described the most interesting details about life in the bush. We studied tracks and smelled various bushes and flowers; we saw the final resting place of "Dan," another male lion who'd recently been killed by a buffalo; we pulled apart dung and termite structures to examine the insides.

We encountered many bizarre beautiful baobab trees, which are definitely my favorite as they always make me think a happy thought about the book, "The Little Prince." They apparently grow from the inside out, and thus become hollow as they age. Perhaps that's part of their spirituality for me: that they grow from the inside out. We were able to peek inside a few of them that had holes in the trunk. In one of these, there was a colony of slit-faced bats living inside - dozens of them flying about. It was unnerving to look in and try to remember that they would NOT fly towards your face; at least, that's what Joe had promised. They didn't; however, it was hard not to flinch anyway.

I also think the various kinds of acacia are attractive, especially the winter thorn and umbrella trees. It's sacrilege for me to like trees, I know, I know, I ONLY like rocks … but for some reason I DO like the trees in Africa.

We saw some stunning birds, including crested francolin and orange-breasted bush shrikes, all the while I was enjoying humming, "In the jungle, the mighty jungle, the lion sleeps tonight …," as we walked single file behind our rifle-toting guide and our Park scout carrying an AK-47.

On our last afternoon at Chiawa we went for a boat ride instead of a game drive, which was fabulously peaceful. We had a leisurely safari floating aboard a flat pontoon boat with chairs set upon the deck, drifting amongst grass-covered islands and small lagoons filled with kingfishers and herons and egrets. On one of the islands a massive bull elephant loafed by the edge of the river, munching on grass. We had so much fun trying to get the boat to drift directly in front of him so that we could take photos of ourselves in front of a huge elephant. Grant was driving our safari that day and he is a photographic perfectionist. He tried over and over again to get the ideal shot, piloting the boat back upstream and drifting down again until the lighting was perfect and the elephant was directly behind our heads. And it worked! We now had our perfect Christmas card photo.

At sundown, we had our Amarula cocktails, taking more perfect Christmas card photo shots of the sun setting over Chirapira Mountain and clinking glasses on the river. After our drinks, we pulled over to the nearby riverbank and hopped out of the boat and into a waiting jeep for our night drive. Almost immediately Grant got a call on the radio from Joe, who said that he'd found a leopard. Grant asked if we were game, no pun intended, to try to get there before the leopard moved, and we all agreed. Grant said to hang on, and we FLEW through the bush! It took about 10 minutes, with Jim and I clinging to the railing with our eyes closed - too much dust for the contact lenses. It was rather thrilling albeit very bumpy.

We spotted the leopard and he was beautiful, with a glistening handsome coat, sinewy back, and lissome figure. He was sitting down and looking away from us. Now and then he turned his head and we could see his striking face. Gorgeous.

After enjoying the leopard's company, Grant continued our night drive and we eventually ended up going down into the dry riverbed of the Chiawa River. Driving along the sand, we could see lights in the distance. "What the heck?" he exclaimed, "Is this some poacher's camp?" However, we could tell he was trying to suppress giggles. "I'm going to CATCH these people!" he shouted as we drove up pell-mell to the front of the lighted area and screeched (if you can screech in sand) to a halt, feigning shock and anger.

In front of us was a magical scene. The camp staff had brought two long tables down to the riverbed and set them up for dinner, complete with white tablecloths and napkins, candles, china and crystal, and decorated with acacia leaves and pods. There was a complete kitchen area, a bar and bar tender, and a blazing fire over which lamb was roasting. Behind were the towering cliffs of the riverbank, with kerosene lanterns nestled in nooks in the cliff. Outlining it all were several fires burning on the sand. To complete the enchantment, there was a "loo" for the ladies, consisting of a wooden seat on a stand (even if it did empty directly on the sand!), surrounded by a white cloth draped on a wooden frame, with a small basin and kerosene lantern inside.

It was an incredible, wonderful, romantic setting. And, as always, we had drinks, then singing by the choir, followed by a tremendously delicious dinner, with stars shining radiantly overhead.

Later, back at our tent we heard a jackal calling … a lonely, longing sound. And also, the haunting cry of the fish eagle. Then, as if to lighten the mood, we heard a hippo "shit helicopter" right by the tent. They spread their dung around over a huge area by frantically twirling their tails as they defecate … wet slapping noises mixed in with whirring. When we awoke the next morning, vervet monkeys were scampering about on our tent roof.

We decided to go for a walking safari the last morning, and Grant offered to be our guide. We boated down the river, and were dropped off at a lovely winter thorn wooded area. The winter thorn is a pioneering species, we learned, as this spot was previously a flood plain until Zambia built a flood control dam in 1955. These now mature trees were the only species that could colonize the sandy loam.

Just in case any of us had forgotten, Grant quietly explained the procedure for a walking safari: "The scout is out in front with his weapon, then I'm next in line with my rifle, and then you are all quietly walking in a single file line behind me … except, of course, when we are in retreat!"

We giggled nervously at the thought, and started off.

It was a bit more unnerving to walk in this thickly wooded area, than in the more open sandy grassland closer to camp. Suddenly, we heard monkey alarm calls through the trees. Grant motioned at us to stop, and we listened intently.

There were impala calling, too, and then francolins starting their shrieking.

"A lion or a leopard!" whispered Grant.

"Let's go find him!"

And we started rushing IN the direction of the alarm calls. Funny, I think I would have been rushing AWAY from them …

We stopped periodically, and still heard the alarm calls. Once, we even thought we heard a lion call.

Or was it a hippo?

As we continued hurrying along, the bird sounds were superb. I heard Cape turtle doves ("work harder, work harder!") and red-eyed doves ("red eyes, what a pity! red eyes, what a pity!"), then the louries with their plaintive call, "Go away! Go away," while groups of babblers babbled, and hadeda ibis began their hysterical laughing.

Then we stopped and listened again. Impala alarm.

Then silence.

Silence still.

Nothing.

Grant turned to us, and "called off" the lion hunt, as he made a U-turn to head back towards camp.

So, now I was in the rear, the closest to the supposed lion. Hmmm. Interesting.

After a few moments, our scout pointed out some tracks - leopard tracks - and at that point we were able to figure out from his spoor that we'd actually JUST missed him. When we'd rushed towards the alarm calls, he'd been walking towards camp, slightly to the left of us. Darn!

We followed the tracks for a while, and that was exciting, but when we lost them in the mulch and got tired of guessing where he might have gone, the sense of urgency was lost and we were back to our walking safari. We saw a buffalo carcass in a grassy hollow; we discussed termites and soil conservation through aeration; Jim found an ant lion for us to examine through binoculars held backwards, providing the effect of magnifying glasses.

Then we heard elephants! Grant suggested that we not sneak up on them as sometimes that backfires, when someone accidentally sneezes, or drops a camera, and the elephants are NOT good to be near, on foot, when they are startled. So we made lots of noise and didn't get too close. We watched several bulls browsing, reaching their long trunks way up into the trees. We saw a mother and a baby, and heard some other elies munching in the thickets nearby. Now we realized we were actually in amongst a breeding herd of elephants. "They're all around!" Grant exclaimed. We had to do a little maneuvering to avoid getting any deeper into the herd, as it's not good to separate mothers and babies.

Out of nowhere, Joe appeared with a vehicle, motioning at us to climb on. Grant was a bit surprised at this interruption of our walk but when Joe brought us to a group of wild dogs, he was forgiven. This was worth the interruption! There were five of them, three lying down and two wandering about, with blood from a kill still visible on their muzzles. These dogs are the rarest large predator in Africa, and are quite endangered. People can come to Africa for years and not see them. We really enjoyed watching them; they are about the size of a medium domestic dog, but with their multi-colored coats and lean bodies they looked like cute little puppies with huge ears. But don't be misled by their cute-ness!

After lunch, Jim offered to help Craig with an electronic issue (something to do with their internet connection … quite difficult out in the bush), and Barbara took me on a tour behind the scenes. It was interesting to see how this operation was managed. She showed me their "store," which has everything they need to serve all these scrumptious meals. It is an unbelievably huge amount of stuff, all kept in a single hangar-like building. Not one but several deep freezers are powered by their generator. Things that just need to be kept cool are kept in a little room lined with walls made of chicken wire, containing charcoal. They wet the charcoal and it acts as a natural air conditioner. As deliveries are made only once a week, there is a lot of figuring for every possible variation of food or drink that a guest might want, not to mention the difficult ones that show up, like vegetarians or people who insist on having only Amarula! I was really interested in the complicated project management that running a camp like this involves.

Barbara also double-checked with me about Jim's birthday cake request. I'd told Bushtracks, our travel agents, that I would like to honor his upcoming 50th birthday on our last day of the trip with a simple cake. They asked what kind of cake was his favorite, and I said, "cheesecake." I didn't realize, of course, that this was an unusual item out in the bush. In fact, no one even knew how to make a cheesecake. So, Barbara and Helen, the hostess at the camp where we'd be on our last night, which was also owned by the Cumings, had apparently been having lots of conversations about this cheesecake. They had to get a recipe, and then order special ingredients that had to be flown in. I hadn't meant for it to be such work but Barbara laughed and assured me it was actually quite fun to solve the problem.

As I was looking at the calendar on Barbara's office wall, I realized that our last night of the trip was August 19. "Hey, our last night is also the full moon, and our anniversary," I mentioned.

"Your anniversary?" Barbara asked. I had to admit that it wasn't actually our ANNiversary, but it was our MONTHiversary. We're still celebrating the 19th of every month, I explained. She smiled.

Then, it was time to move on to our next camp, Chongwe River Camp. It was difficult to say goodbye. We'd really fallen in love with Chiawa and its people. A boat collected us and we were driven upstream.

The Chongwe River Camp is situated right outside the western border of the Park, on a small side river channel, the Chongwe River. As we pulled up to their sandy shore, we were already enchanted. The camp has a large central grassy lawn extending down to the quiet river, with a great view of the opposite marshy bank where multiple shorebirds, hippos, crocodiles and baboons would spend the day. At the top of the lawn is the gathering area and an open fire circle surrounded by chairs overlooking the bank of the river. A long outdoor table for meals, two open-sided thatched buildings with lovely yellow-colored concrete floors, one for relaxing and reading and tea/coffee, the other enclosing the bar, completed the picture. We were greeted by Claire, one of the managers, and Boet, the owner. We sat down in the bar to enjoy a Mosi and were surprised to notice that there was a titanic elephant right behind the building! He was wandering around amongst the large trees on the lawn, picking up the acacia seed pods and eating them. He came closer and closer as we sat there, eventually even reaching up over the roof of the building right next to where I was sitting. Here was an elephant not less than 5 FEET AWAY! I was deliriously happy and not even a touch scared (of course, I was IN a structure). Claire mentioned that we'd have to wait a bit to go to our chalet, since there was SOMEthing in the way …

No problem, we had a wonderful time chatting with Boet and his daughter Karen who was visiting for a few days. Boet started this camp about 12 years ago, and is obviously completely enjoying himself. He is very dedicated to conservation (you can walk to the CLZ buildings from this camp), and has lots of new projects in mind, now that the camp's chalets are completed.

Once the elies moved a safe distance away, Boet gave us a complete tour. Chalets number two to nine are lined up along the riverfront, going upstream from the main buildings. Chalet number one is in the opposite direction - the honeymoon tent - more on that later. There's another little chalet for reading/relaxing and several little thatched loos to visit if you don't want to travel all the way to your tent. A shallow pool that is filled with river water comes in handy in October – aka "suicide month" - when it's far warmer than when we were visiting. Jim declared more than once that at the time of year we were traveling here the weather ranged "from perfect to awesome, and back to perfect."

Boet explained that during those hot October days guests can sit in the pool and drink a beer, while an elephant drinks from the other side of the pool! As we walked, he told us his plans for the future, including an area that he thinks would be a good place for some game viewing blinds, an elephant mud-bath that he's in the middle of developing, and the new self-contained four-bedroom chalet that is being built upstream in collaboration with Robin Pope Safaris. He told us that he has about 35 staff, not that the camp needs that many, but because he wants to provide as many jobs as possible for the local villages. Yes!

At the end of our tour, he brought us to our honeymoon tent. Once again, we realized how wonderfully spoiled we were to have an excellent travel agent planning every little perfect touch for our trip. As I mentioned, this chalet is away from the others, a little downstream from the main buildings. It is right on the edge of the riverbank, nestled in amongst a group of thorn trees. Again the major room is a safari tent, but this one is mounted on a platform of red-colored concrete, with a porch out front and the bathroom behind also made of concrete. It feels wonderful on bare feet! Smooth and pretty. Stones are embedded in the concrete in the bathroom floor and on the short walls around the bottom of the shower pan. Even the bathtub is made from concrete and decorated with stones. The enclosure walls are about 7 feet high, of thatch, and everything is similarly covered with thatch: the shower pipes, the on-demand water heater, the tank of the toilet. Once again, an open-air ceiling so that we could do birding – and elephant watching – while showering or brushing our teeth.

We partook in a bath one evening in our delightful tub. Claire had the guys fill it while we were at dinner, and we arrived to our kerosene-lantern lit bathroom with a full scented bubble bath, lavender incense burning, and an opened bottle of wine with glasses at the ready.

After settling in our room, we rejoined the gang at the fireside for the ubiquitous afternoon tea/coffee and cake. This camp follows a similar routine to our previous one: 6am wake up, 6:30am breakfast at the fire circle, 7 - 11am morning activity (with cake break halfway through), 12:00pm lunch, siesta until 3:30pm, cake break, 4 – 6pm afternoon activity, sundowners, night drive, 8:00pm dinner.

Claire came by and asked us what activity we'd like to do. Karen mentioned that she and her dad were going to go out in a jeep with Claire and her beau Jody, one of the guides, to try to find a kudu that had been killed recently and apparently had IMMENSE horns.

"We want to do that!" we exclaimed. Jody explained that we were quite welcome to join them, but it wasn't really an "activity" per se as we were only going for a short walk to look for the kudu. We explained that we didn't mind; we just wanted to be out in the bush and didn't need such attention as a full activity. Boet then chimed in that we were going to be at this camp for 6 days, so we were going to be practically family and we didn't really need to be treated as tourists anyway.

So it was settled, and we all hopped on a jeep. After driving around for a bit, and having several radio calls with John, the guide who'd spotted the kudu kill ("No, not THAT enormous fig tree, the OTHER enormous fig tree, by the sandy hill!"), Jody finally decided we were close, and stopped the jeep. We got out, and Jody loaded his .458 caliber large game rifle. We spent about half an hour wandering around - I'm sure Jody was marching along an intentionally directed hike, but it felt like wandering around to me, as all the bushy thickets looked the same. About that time, the girls agreed that they were bored and wanted to go see the lions as apparently John had seen some lions nearby earlier. Jody agreed, and when we returned to the jeep he zoomed off in the direction of the lion sighting.

By the time we found them it was almost dark. There were three lionesses and three cubs, all of them sitting on the grass except one large female who was standing alert and appeared to be hunting. We turned off the lights and sat in the semi-darkness, listening. We opened the cooler and had our drinks out in the night air, and sat quietly chatting. It was marvelous! We heard repeated impala and baboon alarm calls, and at one point watched excitedly as a bushbuck sneaked RIGHT BY the lioness as she had her head turned the other way. One very lucky antelope.

Boet spoke with incredible passion as he told us about the death of a beloved elie, "Big Boy," a few years back. Boet and his wife were in camp when they heard poachers' gunshots. They immediately rushed out towards the sound of the shots, and tracked the elephants. With tears he described finding the killed elephant, trunk severed and one tusk already removed. At one point the poachers returned and surprised them with an ambush. Boet and the scouts dove behind the dead elephant and so were protected, but were splattered with the elephant's blood from the bullets as they were shot at while they hid. Luckily, with the arrival of more game scouts as back up they eventually managed to alter the situation, capturing the poachers and also the ivory. But the magnificent creature was dead.

This camp also had delectable food at every meal, including special vegetarian dishes just for me. We'd sit at the table till late at night, chatting and drinking. The bar was also a great place to congregate; before dinner we had drinks and hors d'oeuvres and tall tales of game safaris ("It was THIS big!" and "It came THAT close!"). We enjoyed the people that work at the camp, Claire and Jody, and also the other guide Garth, and managers Lynsey and Maritza. We were also surprised to connect with one of the other families at the camp - two people who went to graduate school with me at Harvard! It truly is a small world.

And each night we listened to hippos gurgling, elephants munching and Egyptian geese quacking – yeah! And I must mention the little frogs making "plinking" noises which was sweet to listen to. On a loo visit in the middle of the night, I could see Orion shining in the sky between the acacia branches.

One day we went for a walking safari with Jody to try to find those elusive kudu horns. Again we were unsuccessful but had a great walk nonetheless. We found a magnificent mahogany forest and sat below a grand old tree with our binoculars and bird books, and identified many new birds. By the end of our visit to Chongwe, our trip total was up to 117 birds, with 27 of them life birds. Jody hadn't taken many walking safaris recently and he told us that it was equally as interesting for him to reconnect with and remember the various birds.

An incident with a porcupine immediately drew blood. I was reaching into the back of Jim's vest, to get out my voluminous Southern Africa bird book, and as I yanked upwards, my hand was stabbed by the porcupine quill on the brim of Jim's hat! It HURT. Jody was quick with the first aid kit and expertly bandaged me. Later that day, we filed off the unbelievably sharp tips of our porcupine quills. It was a relief to find out that the guides are truly well equipped for disasters small and large.

When we later returned to our tent, I discovered that my plastic hair conditioner bottle had been bitten! I guess the baboons were making themselves at home in our bathroom during our absence.

One afternoon Jim was showing some of the guides the sweet photos he'd taken of the red-breasted twinspots we'd seen hopping around at Chiawa. They are a wonderfully beautiful tiny spotted bird, and some of the staff members showed a great deal of interest. As it transpired, they were not as interested in the actual bird pictures as in the fact that Jim could print photos on site. They don't really have any way to obtain a printed photo of themselves, as it would be too expensive and complicated for them to arrange. I expect that it's hard on them to have so many guests using so much expensive camera equipment, and yet perhaps they never receive photos as souvenirs of their dedicated work at the camp. Sometimes a guest would take a photo of a staff member, saying, "I'll send you a copy," but as we understood, the guest rarely did send one after they returned home.

But now, Jim could actually GIVE them a photo of themselves. We offered to take photos of anyone who wanted one, and much excitement ensued. These guys love their jobs, are very proud of them, and it shows. When we took photos of the fellow who made our tea and coffee, he wanted to be posing WITH the tea and coffee. When we gave him the printed photo, he almost cried. He was so grateful that he now had something to take home to show his wife and children.

We were thrilled by their happiness and we did this at each camp on the rest of our trip. What a wonderful gift to give to these people who work so hard and so diligently to make our vacations wonderful.

We went for many game drives at Chongwe, and thoroughly enjoyed them. John drove us and was remarkably informed as well as successful at finding elephants for us. On one occasion, we had a lot of young bulls trying to protect their herds from us. They would make mock charges at us, ears flapping and heads shaking and trunks waving. One young guy really struck a defiant pose with his front foot placed up on a little stump as he shook his head. We could essentially HEAR his defiant " … and STAY away!"

We also drove into the middle of a breeding herd of elephants that were preoccupied with eating so didn't seem to be bothered by our being there. I loved getting the chance to see the baby elephants close up as usually they are completely surrounded by adults. Sometimes they are even encircled if the elies are really worried. Typically, whenever we'd come across mothers and babies they would rush off into the bush. But this time we got to see some really small ones eating as best as they could, without having the expert use of their trunks yet.

We saw lots of majestic kudu, including two standing on top of a termite mound which was quite a sight, some small herds of zebra with their brilliant black and white stripes, dwarf mongooses, hyena, and side-striped jackals.

One afternoon we came to a lovely little glade by a small stream with soft green grass growing all along the shore. The afternoon light slanted through the nearby trees while egrets daintily picked their way about. Jim said he didn't care if there were any animals there, he'd be happy just to sit and absorb this idyllic scene.

As if to draw us back to reality, just a short drive further along the grass we saw lions – directly in front of us. A whole group of them, four females, one young male, and three cubs - incredible. They were clearly eating a meal of some sort. We eventually identified the victim as a warthog from the hooves that we could see dangling from the legs and hanging from the cubs' mouths! Even with the sound of crunching bones, and the snarling between cubs and mother as they tried to grab some more meat, it was a peaceful scene to observe. We sat for some time in the afternoon serenity, watching as the animals finished their meal. For me, the cubs were the most wonderful to see, so very adorable that I just wanted to grab one! The young male was very territorial; I loved hearing his deep rumbling erupting into yelps as one of the cubs would get too near to his food. It was funny to see one of the lions smack another one.

On another afternoon we approached a similar small glade on a stream where there was a group of elephants. The family consisted of two females, an old bull, a tiny baby, and one adolescent male. What a delight to watch them splashing in the mud and the water. The baby was fully visible and while two females flanked him constantly, they weren't so close that we couldn't see him. The young bull spent a lot of time posturing in our direction, but the rest of them frolicked in the water and paid us little notice. After some time, they crossed the river. It was interesting to see the old bull standing guard behind the group as they walked. The water was deep enough that the baby was completely submerged at one point. It was SO marvelous to see his little trunk sticking out of the water, a tiny periscope! On the other bank of the stream they took their sand baths, each elie digging up the sand with a foot, in order to have a pile of soft sand to throw over his back. The baby was in heaven, throwing himself down on the soft ground, rolling around in the sand and simply smashing his face and trunk in it all.

After a thorough romp, the baby stood up and began nursing. His mother placed her right front foot further out in front, to make it easier for him to get to her breast. Before nursing, we noticed that the mother turned sideways to us so that she could keep us in sight, and the other female moved into a protective position. Very tranquil.

An afternoon boat safari was a peaceful and beautiful way to spend yet another idyllic afternoon. We set out with a nice group of guests from the UK (Chris, his wife Annie, and his sister Judy). We pulled the boat up to a small grassy island, and a bird paradise with fantastic sightings. First we saw a black egret – a bird that we've wanted to see for years. He began FISHING. This is a sight that we'd heard about and really wanted to see. He stands in shallow water and suddenly opens up his wings to a full spread, while simultaneously bending over and pulling the wings over his head till the tips touch. It's hard to describe, but he ends up looking like an umbrella on top of the water. The fish below think that this is an overhang on a riverbank, and they crowd in for safety. But, HA! The egret is able to skewer them and have a tasty meal. Not only the egret, but a gorgeous African spoonbill decided to benefit from the egret's ploy, and often snuck his bill under the umbrella to pull out a prize. It was fascinating to watch. Our guide, George, tried to point out some other interesting birds, but all we wanted to watch was the black egret.

After a while, though, the egret delicately walked away, showing us that he is completely black all over, except for brilliant orange socks and feet.

We then drove upriver a little further and pulled up at a grassy marshy spot right off the Zimbabwe shoreline. This was an astoundingly fecund spot. Up on the river bank there were myriads of eland, baboon, impala, waterbuck, elephants, hippos, and crocodiles. In the grasses we saw ground hornbills, black-winged stilts, wood sandpipers, herons of all kinds (grey, squacco, goliath, and little green), jacana, great egrets, open-billed and saddle-billed storks, white-crowned and black-smith plovers, and malachite kingfishers. We even saw two painted snipe, a bird we'd never seen before, with brilliant colors like a harlequin duck. Splendid! We gazed a long time, as the sun got low and the grasses began to glow. Watching the sunset over our beloved escarpment, we sipped our sundowners out on the boat.

Out on the water again one other day, taking a canoe ride, we discovered that it was a bit scarier being so low down in the water, with buffalo or hippo standing over us on the banks of the stream. It was fun, however, to be paddling in the same canoe with Jim while our guide John paddled his own. We spotted several mammoth crocodiles - my nemesis. If there were one, I don't think I could ever be a member of the crocodile fan club. I find them quite evil, perhaps because they have an indiscriminate, cold-blooded killer look. We pulled up for a break at a grassy area, walked up to a small rise, and discovered that we were at the edge of a very bleak plain. It was nothing but hard packed dirt, the site of a previous village which was re-located back in the 1940s. The land remains completely decimated as a result of human defoliation. It was very sad to see that it has been unable to recover from human habitation … makes me worry about our planet itself.

Speaking of ecology, we went to visit CLZ one afternoon. Lynsey walked over with us and gave us a tour. I got chills up my spine … THIS is what I want to do! I want to come back and work at CLZ. We were both really impressed. Just to backtrack quickly, CLZ is the organization that is responsible for training the anti-poaching scouts for the National Park. They are responsible for overseeing game in the Park and make recommendations about how best to protect them. More recently, they've set up a community education program which is fantastic. They realize that the only way to really stop poaching is to teach the new generation of children that the protection of animals is more important than their destruction. Their message essentially is that you may have an economic windfall from killing an elephant and selling its meat and its ivory, but that you'd be better off in the long run economically if you protect the elephant so that tourists can pay money to see it.

Although the economic argument is necessary, it's also important to teach these children the simple glory of wild animals. Amazing as it may seem, kids growing up in villages in Zambia don't ever see wild animals, as they've all been hunted out of the areas where the villages are located. And, obviously the kids don't have the money to be able to visit the National Parks. The school at CLZ was built through the support of the Danish government. It's on a lovely setting on the river bank and is much like a safari camp. It has a classroom, kitchen, dining room, dormitories and bathrooms for the kids, and a few cabins for teachers or visiting parents. Contests are held in local schools for children to "win" a chance to come to CLZ for a two week period, all expenses paid.

The classroom is BRILLIANTLY designed, better even than the science classrooms at the Lawrence Hall of Science in Berkeley, California where I taught science to children back when I was in college. The school teaches the children about the wild animals in their country, the kinds of foliage and seed pods, what the native indigenous peoples used for natural medicines, and a host of other information to help generate pride in their country. It was really inspirational and exactly the kind of thing that Jim and I want to do when we retire! ☺

Back at camp, we watched the bull elies crossing the river to visit. We had walked down towards the spot where they typically crossed and stood on a porch where we could watch them climb up the bank and pass by directly in front of us. It was breathtaking being within an arm's length of a huge elephant. I think I will never tire of watching their lumbering walk, their deep set serious eyes gazing at us as they pass. We had lots of fun following them around, videotaping from another guests' bathroom, as we clambered about trying to get a better glimpse. As an aside, I may as well mention that in fact, the bathroom had an active termite mound as part of the sink structure which was really cool! It was captivating to watch the elephants picking up acacia seed pods. Their trunk tips are like little gloved hands, very precise and capable. And yet those same appendages can suck up nine liters of water to transfer to their mouths. Wow!

One afternoon I learned a lesson from bush etiquette (bush survival?) 101. We were about to leave on a driving safari with a few other guests. I asked to stop at the little loo near the main buildings before walking out to the waiting jeep. While in there, I remembered that I had my friend's camera battery which Jim had charged for her with his solar panel during siesta. I wanted her to have the battery for her boating safari that afternoon so as I exited the loo, I looked for someone to whom I could give the battery. I saw two staff members along the path, and had a complicated discussion with them (they didn't know who she was; I didn't remember her tent number and tried to describe her). While I was talking I glanced over my right shoulder towards the jeep and saw that everyone was on board and waiting, even the driver and spotter.

I handed over the battery and asked the two staff to just TRY to find this person and set off purposefully for the jeep. Hurrying along, I threw sheepish looks towards the jeep, making little noises to indicate that I was very sorry to be holding them up. I even upped my speed a little since I was imagining everyone was becoming impatient to leave. I guess I actually RAN a little.

Well, it's NOT OK to run in the bush. If you run, you will be chased. It's that simple.

Therefore, I guess I shouldn't have started to run.

But I wasn't out in the bush, I was in camp, right?

And there are no wild animals in camp, right?

At least, I had been LOOKING AROUND for potential animals, right? Well.

No, I hadn't been looking.

And as you know, there WERE wild animals in camp.

In fact, one of the bull elephants, the one with only one tusk who was known to be the "ornery" fellow, was standing right next to our tent.

When I started to run, he started to run after me. Apparently he gave quite a mock charge after me. I'm really grateful I didn't see it.

However, Jim did, as did the staff members. They yelled and clapped at him, and he stopped his charge.

When I got to the jeep and Jim told me what had happened, it took me a few hours to recover. I was really shaken. I don't think I'll ever let down my guard again.

One morning at Chongwe we decided to be decadent and sleep longer than the 6am wake up call. Lynsey referred to it as having a little "lie-in" which was funny as it sounded like a little "lion" to our American ears. It was glorious to watch the sunrise over the trees across the river, right from bed, without even having to lift my head. The relaxing morning continued as we just milled around enjoying birding in the trees and watching the baboons, impala and hippos across the way.

Jody called for us around 11:30am, to take us to a secret rendezvous. We had to be "in" on the secret since we hadn't done our morning activity. The staff had set up a complete brunch in the bush, upstream a few miles, by the side of the river. The folks on the game drive vehicle didn't know about it until they drove into the clearing and saw a fully laid table complete with white tablecloth and napkins folded into the shape of cranes! A fire was burning and food cooking. Boy, this was really roughing it.

To get there, Jody took us in one of the boats. He asked if we wanted to go slow or fast. "As fast as you can go!" we responded.

He smiled. "OK," he said, and pushed the throttle. He glanced back to watch the engine moving into place, adjusted the trim, and mentioned "We just have to watch that we don't run over any hippo."

Jim countered, "Like THAT one?" pointing straight ahead. Jody whipped his head around just in time to see the bubbles from the submerging hippo directly in the path of the speeding boat. He immediately yanked the wheel hard over and the boat canted out sideways. We didn't feel a bump so his quick action prevented the engine from hitting the hippo. That was exciting!

We arrived at the riverbank quite thrilled from our little ride and enjoyed a lovely meal with our UK birding friends and Jody and George, while Maritza and several staff members served our meal.

Our last night at Chongwe we listened to hyena calling during the night, a tragic and mournful sound. We slept in until 7am, waking to the awe of the sunrise and the multitude of birds in their dawn chorus.

Once again, it was hard to say goodbye. We really had begun to feel like family and didn't want to leave. Jody was to take us on our canoe ride to the next camp, Sausage Tree Camp, and we insisted that he bring Claire too. We didn't want him to have to be all alone in his canoe. It was fun having a "double date" as we canoed down the Zambezi. The boys did a bit of fishing and Jody caught a large tiger fish. Jim got several bites but they managed to sneak off the line every time.

We took a little side channel for part of the trip, and again enjoyed seeing animals along the bank. At one point we saw a buffalo crossing the stream and we had to stop from getting too close by thrusting our paddles down hard into the muddy river bottom. Luckily at that place the channel was quite shallow. I was videotaping the buffalo's progress and continued to as we passed him by. He was standing quite close to the edge of the water staring at us. Just as I turned off the video he decided to charge us! It was rather frightening. He towered over us as we were sitting in the canoe, but only came a little ways towards us before backing off. Guess he was all bluster - phew.

We stopped for snacks along the way on a little promontory over the Zambezi. Jody made the mistake of telling us a few scary canoe stories BEFORE the end of the canoe trip. Like the time a buffalo charged Jody and actually got to him. Jody hid under the overturned canoe. And he recalled the time a hippo BIT OFF the front end of his canoe - luckily there was no one in the front seat.

I was a little more nervous as we then continued down the main river but it didn't detract from the joy of seeing elephants on the river bank and watching a yellow-billed stork wiggling its toes in the water to try to catch fish (which think his toes are worms). When we got near the entrance to the channel that leads to Sausage Tree, Jody paddled over to our canoe and told us how we were going to manage it. The problem with the channel entrance is that there are always two pods of numerous hippos on either side, standing on the banks of the river grazing the grass. As we got closer they'd get nervous and jump INTO the water. The same water that WE were in. Not too bright, are they? I mean, if they're scared of us, shouldn't they just stay OUT of the water? But hippos only feel comfortable when they are in deep water.

Jody knew they were going to jump in, and he wanted us to know too, so that we wouldn't panic. He also wanted us to be sure to stay very close to his canoe, and to follow his instructions exactly.

All of this serious discussion was stirring up my nervousness – again!

Jody said that I should videotape the hundreds of hippos leaping into the water, explaining that it would be quite an amazing sight. I was concerned that if I did, Jim would have to do all the paddling, but Jody guffawed, "Aw, he'll be able to manage it just fine!"

So, as we came to the channel entrance I had the video running, with Jim behind me doing all the paddling. As expected, the hippos leaped in, one after the other, and as expected, it was quite a sight.

Then suddenly, one of the hippos apparently decided to get in front of our canoes - this was getting to be a habit with us and Jody. Jody yelled, "Back paddle!" Jim started doing just that but we were still moving forwards. "Back paddle!" Jody yelled again.

Then Jim yelled, "DROP THE VIDEO CAMERA!" I dropped it in my lap, grabbed my paddle, and began back paddling. And this worked. We stopped our forward progress, the hippo moved on, and we continued along our way.

Of course, the video camera was running that whole time, so it's a rather fun scene … the hippos, the yelling, then a sudden flip and the picture is sideways, watching my paddle moving frantically.

After that excitement, it was a peaceful relief to maneuver calmly through the grasses of the lagoon towards our camp. Just around a corner, there it was, right on the bank of the river. "All" we had to do was cross the main channel from the little waterway that we were in. There was only the ONE hippo in the way … or were there two? As we got closer, a second hippo suddenly appeared, but luckily we were on the opposite bank from where he climbed out. The first hippo we recognized as the "local" hippo that has been hanging around the camp for years, named Frank. His new sidekick is a more recent friend and has been named Stein. Sausage Tree can now boast of their Frank 'n Stein.

We canoed over to the shore, pulled up onto the sand, and were welcomed with lovely cold towels. We met Jason, one of the directors of the camp and Tammy, the assistant manager. Again, we realized that even though we had hated leaving Chongwe, we were going to love Sausage Tree.

The main sitenge at Sausage Tree is quite open as at the other camps. But this sitenge is built with smooth orange-colored concrete floors that flow into benches and down steps to the grass. In front of the sitenge is a beautiful deck over the edge of the riverbank, with tables and chairs and standing umbrellas. There is even a large section of deck devoted to the "fire circle," including a submerged circular concrete disc for the fire at night. The dinner table is set in the sitenge under the roof but can be placed out on the deck under the stars, depending on the temperature. There is an elegant guest bathroom (also with the decorative wooden guinea fowl family which we'd admired previously). There are several chalets along the river going upstream, and two honeymoon tents in the other direction, separated from each other and more remote.

We had a nice lunch with Jason and Tammy, including Jody and Claire before they left to go back to Chongwe, and then we were taken to our chalet. Once again, we were given a honeymoon tent. We have been completely spoiled.

As mentioned in our previous trip, the tents at this camp are unique: instead of being the typical square canvas safari tents, they are round, with walls of thatch surmounted with a white canvas roof suspended from a pole in the middle of the room. The portion of the tent that faces the water has an opening in the thatch, with a comfortable sofa positioned so that you can sit and look directly out and down onto the water. This opening is covered with canvas at night, as creatures could otherwise wander in. The floor of smooth colored concrete extends out to the bathroom, which is also constructed from smooth concrete formed into 6 foot high walls. A glorious outdoor shower and sink, and a loo built right into the concrete bench completes the bathroom. Since this is a honeymoon tent, there is a delightful tub where we later took another great bubble bath – here on safari, in Africa! The bedroom is spacious, with an immense bed surrounded by a canopy of mosquito netting, tables, chairs, a dresser, and a dressing table with armoire. Very decadent-feeling.

As at the other camps, we had a 6am wake up call, but at Sausage Tree each chalet has its own "valet," who comes in the room with a tray of tea and coffee to help get the day off to a proper start. Simon was very sweet, and even though we couldn't actually see him without our contact lenses, we always tried to be gracious when he'd waken us from a deep sleep. The rest of the schedule was the same, with morning activities, a light lunch, siesta, cake, afternoon activity, sundowners, night drive, and dinner. We especially remembered the chef, Honore, from our previous visit; he is originally from the Congo but trained in Paris. Incredible food! We were at this camp for four days, and loved it just as much a second time.

For our first afternoon activity we opted for a game drive. When we walked out to our vehicle, there was Moses. We remembered him from our trip two years ago! It was great fun to reminisce and we were flattered that he remembered us. He reminded us of the time that we took a canoe ride together, and he tested me on my bird knowledge as we set off. We had many great game drives with Moses and most of the time, with just the two of us as guests. Our first night out, we spent a lot of time sitting quietly, listening to the African sounds, and enjoying being out in the bush. We really FELT Africa this way, being slow and quiet. We watched lots of elephants ... saw a baby elie suckling ... and an entire gaggle of sacred ibis roosting in a tree. We stopped for Amarula sundowners by the river bank, with a perfect sunset over Chirapira Mountain and the escarpment lining the horizon.

While we were at sundowners, we noticed another vehicle nearby and our spotter went to chat with them. They mentioned that they'd seen a leopard in a certain large thicket and so we took off on an extended leopard hunt. It was a lot of fun ... Moses caught a flash of the leopard's backside at one point ... but we never saw more. He was really hidden in the deep brush. But it was certainly adventuresome to conduct the search - at one point while we were quietly straining our ears for leopard calls, we heard a lion.

We gave up on Mr. Leopard and drove out onto an open plain. Jim whispered, "This is the kind of place you'd expect to find a porcupine," and a few minutes later, we saw one. We feel we are really starting to understand this place. We bumped into another vehicle, from Kasaka, and heard Maritza's voice say, "Hi, Jim and Cynthia," out of the dark. It was a great honor to be recognized – we felt like locals. On the rest of the trip, we often bumped into vehicles from either Chongwe or Chiawa, and it was always fun to recognize the guides – and to be recognized.

Later, we found two genets. I say "we" found, but actually our spotter found them, at great distance. We noticed that all our guides and spotters, at all the camps, have miraculous vision!

During the night, we not only got to hear the "normal" sounds of hippo, but we also heard a loud "whooshing" sound ... almost like waves on a beach ... that turned out to be a hippo EATING grasses in the water. During a loo break in the early morning hours, I was fortunate to catch sight of the very red moon sinking down towards the horizon.

On our other game drives we tracked lions, though we didn't actually see them, watched hyena, and saw two honey badgers clambering about together in a tree. We had a superb time watching a troop of baboons with several babies playing together. There was a lot of mutual grooming going on; one mother was being groomed even while she was nursing. At one point she apparently tired of the whole affair and started to walk off with the baby hanging from her nipple, trying to run alongside to keep up.

We spent lots of time trying to spot birds. One time we heard a honey-guide calling and it took us FOREVER to find it amongst the foliage in the tall trees. Moses imitated its call, it called back to Moses over and over, yet it STILL took us a long time to locate him!

Moses taught us another name for the star chestnut tree. I already knew it was called a "false baobab," but he explained that it is also called the "tick tree," as its little grey seed pods look like swollen ticks. Moses found a few to illustrate.

One morning, we saw a young leopard right near camp, relaxing on a sandy hill. He was really beautiful, almost two-toned with darker spots on his face and shoulders, and lighter below. He didn't hang around long though, bolting off when we stopped to look at him.

We also came across a vast breeding herd of buffalo and sat right in the midst of them for a long time. So many amazing individuals from tiny babies to sturdy old bulls.

On one game drive, there were several other people in the vehicle, all interested in birds. It was quite amusing to spend a long time trying to identify a "little brown job," or LBJ as they are often referred to, way up in a tree. Moses was entertained by watching us discussing the possibilities, poring over our various bird books, and deliberating at length. We finally agreed it was a green-capped eremomela, but it was not an easy decision. Luckily, Moses agreed. By the end of our stay at Sausage Tree, we'd increased our trip total of bird species to 134, with 32 being life birds.

On another game drive, I decided to sit up front as the vehicle was rather crowded. It was great fun practicing (pretending?) to be a guide. We came across an unbelievable sight at a "dambo" - lagoon - that was drying up. There was very little water, nonetheless, catfish still wriggled about in the mud making it a bonanza treat for storks, both marabou (which look like funeral undertakers) and yellow-billed; there were at least 20 of each. Just as many fish eagles roosted in a nearby tree, calling and calling with their plaintive shrieking cry. They would swoop down periodically to grab a fish. It was amazing to watch the birds flipping the huge fish around with their beaks, to get them turned into a direction so that they could fit the fish into their mouths and then slide down their gullets!

During one of our afternoon siestas, Jim printed out photos of the staff. I enjoyed taking the pictures and they were greatly appreciative of this gift. Some guys wanted their photos taken in the bar, standing next to a beautiful wooden bird structure; others wanted to be out on the deck, relaxing while overlooking the river. They all got quite a kick out of being given the photos, and enjoyed looking at each other's pictures and teasing one another.

One afternoon, Jane arrived. She is the manager of the camp and had been away in Lusaka for a few days. We had a funny encounter during lunch. I had mentioned to one of the other guests that we were celebrating Jim's 50th birthday, and Jane spoke up.

"Oh! YOU'RE the cheesecake!"

I was flattered, thinking that she meant that if my husband was FIFTY, I must be the young tart that he'd married.

"Oh, really, there isn't THAT much of an age difference," I demurred, humbly glancing downwards and smiling shyly to hide my pleasure at being considered "eye candy."

"What?" Jane queried. "I was thinking that it was YOU who had ordered the cheesecake for the birthday celebration …"

Oh, right. THAT'S what she meant.

It was rather embarrassing to then explain what I'd THOUGHT she'd meant by her comment. Such delusions of grandeur!

One night, when we returned from our night drive, I was given a "gift" from an exuberant Tammy, an envelope with the word "mom" written on it. Enclosed was a printout of an email from Jessica! I'd been getting more and more lonesome for the kids, and Tammy had offered that I could send an email that afternoon and this was the response. I'd just wanted to make sure that everyone was ok. Jessica said that she'd talked to Mark, and he was doing fine, and that her wedding plans were all going well. I heaved a huge sigh of relief.

One afternoon we had a really special encounter. Moses had heard that Douglas was recently seen in the dry Chiawa River bed and he suggested that we try to find him. We agreed with the plan, and off we set. We arrived at the river bed and began helping with the search … looking about from side to side, scanning the view, bending over and looking for tracks. At one point, Moses had stopped the vehicle and while I happened to be bent over the side of the jeep staring at some tracks in the sand, I heard one of the other two ladies on board whisper, "He sure looks lazy!" I wondered what she could possibly be talking about. I sat up and looked over the other direction to see that Douglas was RIGHT THERE. We'd driven up to a male lion, circled around him, and parked the vehicle, before I even realized it was Douglas whom we'd found. Guess I'm not ready to be a guide just yet.

He was lying on a cushion of grass, sound asleep. We hoped that he'd wake up so we could see his mane and get the full glory of the male lion, but he just lay there most contentedly.

After a few minutes, it actually became a bit boring.

Moses asked if we'd like to go down the riverbed a little ways for sundowners, explaining that once it got dark, Douglas would probably wake up and call to his females. We agreed this sounded like a great plan. We drove down the sandy riverbed a bit, then stopped and hopped out. I noticed that this evening Moses didn't set up a table and tablecloth for the sundowners; also no one asked to go to the loo.

Were we all thinking about the LION just a little way up the river?

We had our drinks and snacks as we stood and chatted for fifteen minutes or so.

Just as it got dark, we heard him.

And it was LOUD!

He called a strident, deep, agonizing groan, followed by many shorter grunts.

"Let's go!" said Moses. We didn't need any more encouragement to jump in the jeep.

We zoomed back up the river bed and parked in front of Douglas again. He was now sitting up, looking in the direction of our jeep but he didn't seem to pay us much notice. We sat and watched him in the growing darkness until he called again.

And this time, it was INDESCRIBABLE.

It was one of those moments in life.

There are no words.

It was in my stomach … it was all over my body.

It was amazing. It simply reverberated.

The sobs leapt out of my throat.

I was in awe. What a primal, gut-wrenching sound.

A few moments later, he got up and walked around a bit, and then stopped. We re-positioned the jeep directly in front of him again, and Moses suggested that he'd turn on the spotlight when the lion called again.

This time, Jim captured it on video. It was even more awesome to SEE the lion in the throes of his call … he was heaving with the effort. This guy called his heart out.

And I simply cried and cried. It was SO emotive! A true encounter.

When the calling stopped, Jim was so excited that he played back the video recording.

Oops.

Douglas looked directly at us, intently listening, and then began walking towards us.

Moses said quickly, "Think we ought to be going now," and started the jeep. We were all in complete agreement with his decision…we didn't want to see what would happen when Douglas met his putative adversary.

When we returned to camp I just had to hug Moses and Tammy, "That was FANTASTIC!"

The energy of the wilderness seemed to continue throughout the night as we heard astonishingly loud hippo calls. In fact, I was so interested in getting the sound on video that I was up at 4am, clambering about the bathroom, quite naked, trying to point the video sound recorder in the direction of the hippo.

The next day after our morning game drive, we were told to dress in shorts and sandals and to report back to the sitenge in half an hour. Wonder what's in store? When we gathered together again, we were all loaded onto boats and driven downstream to a complete luncheon IN the river. A full table was set with umbrella for shade, ready and waiting for us. It was marvelously fun to wade across the sand bar to our table, and to sit in chairs in the water with our feet immersed in the river while we ate.

Jim and I collected some sand from the sparkly river bed to bring home for our sand collection.

I managed to squeeze in at least one nice relaxing moment when we returned to our tent. The guys had filled the tub with a lovely bubble bath. The minute I climbed out of the tub, we had another elephant encounter. A towering bull elie crossed the river right in front of out tent, and then climbed up the river bank and walked by the walls of our bathroom. There I was again, naked, clambering about the bathroom, and filming away.

The next morning, we had one last game drive to see elephants and we did see a breeding herd so I was once again able to indulge my love of seeing the babies. We also tried again to find the lionesses but no luck with the kitties.

Before long it was time to leave Sausage Tree for our next camp, and once again, it was very hard to say goodbye.

We were driven about an hour down the Zambezi River on a boat, with a gentle breeze and calm water accompanying us. We passed many pods of hippos, those on the banks of the river jumping in, and those in the water quickly submerging when we got close. It looked like a "whack-a-mole" game! There were also many elephants; one large breeding herd was at the water's edge but as we approached they became nervous and moved back into the tall grasses, forming a protective circle around their youngest baby.

We pulled up in front of our next camp, Old Mondoro Bush Camp, and tied up at the sandy riverbank. We were met by a staff member with refreshing cold towels, and also by Helen, the camp manager. This camp is different from the others in that it's more remote, and has fewer amenities. It's a true old-style "bush camp." I fell in love with it immediately. The main sitenge is quite simple; just one smooth flat colored concrete floor with a canvas roof supported by tall wooden posts. Inside are a sitting and a dining area, with a small table for a bar and another one for serving food. Very basic, but very sweet. A bathroom conveniently located near the sitenge means that you don't have to walk all the way to your tent; it is made of thatch and built into a tree on the bank of the river. There are only a few walls of thatch; the side facing the river is quite open, so that you can do some hippo spotting while on the loo!

There are only four chalets in this camp, and for the first two nights we were the only guests; another couple arrived for our last night. Helen said that it was usually full so this was quite a relaxing treat for her. And it was great for us! We got to spend a lot of time with Helen and her husband Roelof, the guide, and really enjoyed their company. They are very serious, interesting people, and wonderful to talk with.

Helen showed us our chalet, and again I cried with joy. It was perfect! The floor was green-colored concrete, molded in the bathroom to a sink and shower basin. The room had thatch walls with canvas windows and door (open in the day, closed at night), while the bathroom also had 6 foot thatch walls but no roof. The shower was built around a tree trunk! The queen sized bed was draped with a mosquito net while small bedside tables with kerosene lanterns completed the picture. The room looked out on a small lagoon off of the Zambezi, bright green with lily pads.

The "schedule" at this camp was the same as the others, except that our morning wake up was slightly different. On our first morning, I heard some unusual noises and I poked Jim.

"It's the buffalo!" We had been told there was one that liked to hang out near our tent.

We listened a little longer. We found it funny, though, because it actually sounded like a wheel barrow.

We listened a little longer, and it sounded even more and more like a wheel barrow. What a weird thought.

A few moments later, I heard the sounds of water splashing … into the canvas bucket in our shower. Ha! It was the staff bringing us hot water to wash our faces. In a wheel barrow.

"Funny buffalo," Jim observed.

We took game drives both morning and afternoon, and Roelof was our guide. He was fantastic, very respectful of the game, and always careful to whisper when we were observing. He would quietly relay detailed information about each of the animals we were watching; he clearly loves all the animals in the area and knows them well.

We watched warthog piglets with their parents. A litter is typically six to eight piglets in December; by August most families have only one to three piglets left, as many creatures find warthog babies to be a tasty snack.

We watched shy, magnificent kudu. We saw large troops of baboons and for the first time some peeked at us from behind tree trunks. And we saw many, many elephants - yeah!

The ecosystem here is much different from the other three camps. It is a more austere environment with almost no brush or thickets, mostly tall trees and very short grass. All this openness was refreshing while making it much easier to spot the game.

Scents of wild sage, potato bush, and wooly paper bush filled the air; we loved these smells.

We had our Amarula sundowners up on a wide open grassy plain, spotted with zebra, in awe of the outsized red globe of the sun set over the escarpment. The moon was almost full, too, and red as well. No more stars could be seen at night now, with the moon so full, just the two planets. At times it seemed as bright as day - this dazzling moon is hard on predators.

Our night drives were exquisite. We saw several genets, and one in particular was quite amusing. We watched this fellow for a long time, as he walked along through the leaves and sandy dirt. Suddenly he stood on his hind legs, like a meerkat! Quite adorable.

We saw an African wild cat several times, and had especially unusual experiences watching honey badgers. In fact, we saw them so frequently that Jim suggested that they should be the mascot of our trip. One night we watched a badger couple building some kind of a den, each one taking turns digging frantically in the sand while the other slept. I would have loved to peek down into the burrow they were digging.

After our night drives, we'd arrive back in camp for sherry or port around the fire by the river, and then would make our way over to the table, also set on the riverbank, for our delicious candle-lit dinners and long leisurely conversations.

During our first dinner, Roelof mentioned that there was a lion somewhere close by that had been snared, and that CLZ had been searching for him with their plane. We said that we'd love to help with the search. Roelof wondered whether we'd mind not having normal "game drives"… needless to say, we thrilled at the opportunity to do something that could make a difference. So we helped search for the lion during several mornings and afternoons, and although I saw lots of "grass lions" and "impala lions" and "log lions," we unfortunately didn't find any actual lions. We heard later that luckily he'd been found and darted so the snare could be removed. The veterinarian thought he was going to recover ok, what a relief. It was truly exciting to be part of the effort to try to help a wounded wild animal.

During our drive, Roelof pointed out that the mahogany trees are all browsed to exactly the same height above the woodland floor as that's as high as a kudu can reach. They are the only antelopes that can eat mahogany. In a somewhat complicated feeding process, they must eat the clay of termite mounds first, as the clay binds to the tannins in the mahogany leaves and protects them from that poison.

We saw several bull elephant duos, with one older and one younger elie, the younger one obediently following the older one around. We watched an elephant use its trunk and teeth to help break a very strong tree branch. And we even saw an African wild cat snoozing. It's very unusual to see them during the day, and we approached quite closely yet still he never woke up. We also enjoyed a frisky group of zebra, and took photos of a gorgeous saddle-billed stork standing very still and bathed in brilliant sunshine.

One time as we were driving along the track we came to an area which was more thickly wooded. As we wound around through the bushes, we unexpectedly came across a rather large group of elephants. All of the elephants we'd seen earlier in the day had been bulls, and so at first I was overjoyed to see some young elies, and in addition a tiny little baby. I must admit I was actually thinking to myself that I'd like to drive in even closer, especially to the small grouping to the left, comprised of three females with the tiny baby, and a large –actually incredibly large – matriarch behind them, when I noticed that the large one had started to walk purposefully towards us.

She wasn't altering course. Her ears were back and her trunk was down.

Roelof started the vehicle almost as soon as we'd come to a stop. He watched the matriarch for just a moment, and then put the car in gear and started driving. The track went towards the group of elephants before turning to the right, but Roelof turned immediately to the right through the low bushes.

And that's when she started her charge.

I didn't know that elephants could run that fast. We were going as fast as we could, winding through the trees and the bushes; for a while she was gaining on us.

I looked back at her, and what an incredible experience! This elephant was out for blood.

As we passed by the second grouping, a female with a small baby also joined the chase, so now we had two angry enormous female elephants, and a baby, running after us at full speed.

It was quite some time before we pulled significantly ahead of them, and they gave up the chase. We stopped the car and all breathed deeply. That was close.

Roelof explained that those females would not have stopped. They were planning to kill us, if they had caught us. This was NOT the usual bluster. Luckily, he knew how to read the animals. He realized as soon as we'd stopped that the matriarch was heading straight our way. He explained that this group had recently come into the area, and he'd noticed that they were particularly anxious about humans … perhaps because the two older females are tuskless. This is a genetic trait that is seen more and more these days, because the animals with tusks have been poached for their ivory. Tuskless animals are more nervous presumably because they don't have tusks to help them fight to protect themselves and their babies.

Luckily, the other two groups of breeding herds of elies that we came across later that day all had their tusks, and were not nervous about our presence. So, we got to watch more elephants, including young ones. I was in heaven. Again. Guess I'll never tire of watching elephants.

We had some really lovely experiences with elephants around the camp, too. Both days while we were relaxing in our chalet during siesta, and I was writing in my journal, several large elephants came down to the little lagoon behind our tent for a drink of water. One day, I watched as one of them walked up to a very thick tree, leaned forward, carefully aligned his trunk with the tree trunk, purposefully splayed his back feet out, and then PUSHED on the tree, shaking and shaking, while the seed pods dropped to the forest floor. Amazing strength. Then, he proceeded to pick up the pods one by one, with the little delicate tips of his trunk demonstrating surprising sensitivity.

Another elephant joined him and they crossed the stream. I watched them take a mud bath, then a sand bath, following which they sauntered into camp to graze from the trees near our tent.

Next day, there were four elephants on the far bank of the stream, and this time we watched as two of them nuzzled each other … trunks entwined, tusks interlocked, heads rubbing together and making low rumbling noises. Jim and I were spellbound, watching this most romantic tender behavior. At one point the larger one turned, and the smaller one leaned into his (we presume "his," as we still weren't very good at distinguishing male from female) back side, rolling her trunk over his back and hugging him with it as she swayed against him, rubbing her face on his side.

Wow.

We continued to watch the elephants from over the top of the thatch of our bathroom while we took wonderful hot showers, the heated water having been hoisted up into our canvas buckets by the staff during lunch.

Our "staff photography" session was easier to do at this camp as there were only seven guys employed. Once again, they loved posing either at their work stations, or relaxing by the side of the river. Some of them even opted for photos in the boat, acting as if they owned the world! I took photos of them as a group … they demonstrated great camaraderie as they draped their arms around each other. I suggested a "funny" picture, and they enjoyed making hilarious faces at each other.

On one afternoon drive, we came across a herd of at least 400 buffalo, walking across our track. We waited and watched, and it was almost eerie as the dust they kicked up silhouetted their shapes against the slanting afternoon sun.

That evening, after sundowners on the grassy plain, we passed by Jeki "International Airport," our private joke as we like to call it that, but it's really just another dirt strip. As we drove along the side of the long dirt runway under the full moon, I was reminded of the times that Kalpana and I would land at airstrips like this, and sleep under the wing of the plane. I cried a little bit. Jim put his arm around me. "She would have loved it here," he quietly whispered.

At the end of the runway, we saw a really unusual sight: a Sharpe's grysbok, a tiny delicate little antelope and very skittish. We just saw him for a moment before he bounded off into a thicket. Even so, it was a very special moment.

We spent a long time examining two different nightjars trying to determine exactly which kind of nightjar they were, referring to multiple different bird books. We concluded that one was a "fiery-necked," and the other was a "freckled." They each sat still in the dirt track and let us evaluate them at length with a bright spotlight. They didn't seem bothered at all.

On our return to camp, while sipping port by the river around a blazing fire, Helen suggested that we go say "hi" to Norman, the hippo that has taken up a nightly residence outside of their chalet. We walked over, and sure enough, there was the hippo! Very sweet that he seems to like human company.

Next day we asked Roelof to take us on a walking safari, and it was, as always, quite fun. We were careful to walk single file; we studied the plants, the termites and the dung, and enjoyed the thrill of peering at some bull elephants from behind a very big and sturdy tree. We also had a fantastic sighting of a malachite kingfisher – a tiny little bird with extremely bright plumage, orange and blue and white. He posed on the tip of a grass stem for us for a long time, so that we could see even the black and blue banding on his head.

Our last day in Africa came around too quickly. We had a wonderful afternoon game drive observing an entire family of dwarf mongooses running in and out of their home in an old termite mound. We could see that one was always "posted" on top as look-out. We watched as they dove into the various entrance holes upon our arrival, and then slowly emerged to return to scrabbling around in the leaves and dirt, looking for goodies.

On the bank of the Chakwenga River, which we crossed during our safari drives, we watched a colossal bull elephant across the river. He stood quietly for a while, and we enjoyed his magnificence. Then he turned, walked across the river, and climbed up the bank, directly in front of us! Roelof quietly tapped on the plastic dashboard, and then more loudly tapped on the metal, in order to alert him to divert his path somewhat, as the elephant appeared to be intending to walk directly ONTO our jeep. He did eventually veer off to the side, and slowly, majestically, lumbered past. It felt like a "good bye" from the elephants.

We came across a large group of white-fronted bee-eaters, gorgeous green, orange, and white birds, taking a sand bath together. There were at least 30 of them, wriggling and flopping about in the sand. They flitted in and out of the holes that they'd dug into the bank of the river, and it was sweet to see their little faces peeking out. There were also a large number of carmine bee-eaters at that same spot, larger birds with even more dramatic plumage. They have brilliant blue heads and backs, while the rest of the bird is a vivid scarlet. When in flight they have a very long racket tail hanging behind. Gorgeous! They are migratory birds, and had just arrived at this spot. We saw about 50 of them, but later in the season there are apparently flocks numbering in the thousands.

Speaking of birds, by the end of our visit to this camp, our total trip bird species sightings was 142, with 35 of them being life birds. That brought our grand total of bird species seen in Africa to 249. We were absolutely thrilled by that. We tried to find one last new species on our morning drive to the airport, to make it a round 250, but no luck. Guess we have to come back …

That night we had more sightings of porcupines, and saw the green eyes of spiders on the ground. It was bizarre to see so many little points of light shining out at us as their eyes reflected the lights from the vehicle. Catching sight again of "our" two honey badgers as they scampered along was the perfect nightcap.

Back in camp, after drinks by the fire, we had another delicious, splendid dinner. As we took our last mouthfuls, the cook and the staff came slowly walking out to our table by the river, singing a beautiful African song and carrying a cake surrounded by candles.

Of course I cried.

I expected them to end their song with, "HAPPY BIRTHDAY, JIM!" but instead they ended with "HAPPY ANNIVERSARY!" and placed the cake – the infamous cheesecake – in front of me. I guess my comment to Barbara about it also being our anniversary was taken seriously! We were so touched.

We proceeded to devour the most delicious cheesecake we've ever had, covered with Amarula-flavored topping!

Our wake up call was at 5:00 the next morning. Helen drove us to Jeki International Airport to start our trip home.

I'm crying again, just writing about it.

We WILL be back…

Chapter 4
Africa 2006

◆

A Year of 4's: 4 years, 4 friends, 4 camps, 4 leopards!
    As usual, we began thinking about our next trip to Africa about 5 minutes after returning home from our previous one. We didn't really get serious until we chatted to our dear friends Ellen and Ron, who had previously mentioned that they would be interested in going to Africa with us someday. They subsequently decided that this year would be it, which was very thrilling news for us.
    Our discussions with Bushtracks began once again, and once again they were invaluable, providing all the help we needed with our planning.
    Jim and I both wanted to go back to the Lower Zambezi, which we'd begun to think of as "our valley," and to Chiawa and Chongwe in particular. We also felt we'd like to go back to the South Luangwa Park, since we didn't visit there last year, and we really wanted to go back to the town of Mfuwe to spend time meeting the people who live and work in that small village. So it looked like Zambia was going to be the country of choice again. The only problem was that we also wanted to revisit Namibia, perhaps next year …
    Once we started organizing the trip we agreed that we'd be together for most of the time and separate at the end, as Ellen and Ron wanted to go further afield to scuba dive and fish. We asked Bushtracks to arrange several camps in the South Luangwa in an area we hadn't explored before, and also Chiawa and Chongwe. We proposed going in late August, as we are becoming progressively braver each year to go when it's hotter and when the animals are more congregated around sources of water. We're still not brave enough to go during the suicide heat month of October however.

Bushtracks set up a wonderful itinerary for us and the ensuing months were spent preparing, with the excitement building daily. Ellen and Ron purchased their kit in collaboration with our input: "Is this the RIGHT shade of green?" "Is this jacket too heavy?" "How many pairs of pants do YOU bring?" We shared our piles of equipment with each other as they grew. On several weekends, as we are neighbors, we could be seen dashing to each other's houses, exchanging ideas and even stuff - we loaned them some duffle bags we no longer use, gave Ellen a binocular harness, and Ellen gave me a "first class seat" blow-up pillow.

As a gift, I bought copies of "The Cry of the Kalahari," and "The Eye of the Elephant," for Ellen and Ron. Ellen then discovered that the authors, Mark and Delia Owens, were actually going to be giving a lecture in San Francisco in June, so as a gift to us, she bought tickets and arranged for the evening event. It turned out to be a wonderful reception and slide show, with lots of tears on our side, and signing of their new book which we all read. A great warm up for our trip.

About a month before we were to leave, Jim and I had become so excited by all the planning that we had actually already completely packed everything. I mean everything! Jim ordered a new case for the solar panels and printer in a more Africa-friendly beige color (last year's was black), and he bought a new large camera lens.

A few weeks later a terrorist threat concerning liquid bombs resulted in a change to the baggage regulations, so we had to re-arrange and re-pack everything to fit with the new rules. We used to carry all our luggage with us; now we'd have to check most of it. Jim ordered a new hard-sided case for our camera gear, and we spent a fun evening together while he figured out new configurations for each lens and piece of equipment and I checked through the entire Southern Africa bird book to re-familiarize myself with all those marvelous birds.

Our preparation for this trip went beyond previous bounds. We spent three nights in a row up in our Africa-decorated guest room while I read out loud the stories of our previous trips, and flipped through our photo albums and sipped Amarula with our African animal sound CD playing in the background.

Ellen came over to cut my hair short, and Jim stopped shaving … and we got more and more excited.

We emailed Grant at Chiawa, asking if we could bring anything. He and Jim corresponded about his needing some new radios with US frequencies to put in the tents so a guest could call for help in case he/she encountered a surprise attack. We managed to pack these extra four radios in our already quite tiny luggage.

I brought our African CDs, including the Chiawa choir, to work and loaded them up on my iTunes, and listened to African music each day while at work. We went shopping for a bathing suit as it would be warmer on this trip, and so we figured we may actually take a dip in one of the plunge pools that we've had to skip in the past. I found one that had a few beads hanging in front, then went to "Global Beads" to get replacement African-looking beads. How's that for commitment! Jim made foot rests for Ellen and me to use on the flight, based on a design idea from our daughter Jessica. They provided a way to hold our feet off the floor, using straps to attach to the tray table, and using the in-flight magazine rolled up to provide a flat surface on which to rest the feet.

As if all that weren't enough, on Sundays we walked to our favorite local restaurant with Ellen and Ron for breakfast and Africa discussions. Ellen told us that Ron had installed a "count-down" calendar on her computer, in honor of the upcoming trip. Jim suggested that we get out our pocketed cloth advent calendar and use it for this purpose too. We each secretly purchased 11 little tiny Africa-related items and filled in the calendar's pockets, representing the days until our departure. Of course, gifts for me were placed in the "odd" numbered pouches … one would think we didn't have anything else going in our lives during this lead-up to our vacation!

One of the items I bought was a wind-up plastic frog for Jim, however he wasn't sure of the connection between a frog and Africa until I reminded him of the little frog that used to jump up out of the drain and onto the floor of our concrete shower at Sausage Tree Camp in the Lower Zambezi. He had to agree with me on that one.

Finally, The Day arrived. Our friend Diana arrived at 10am to take our traditional "before" photos which she's done for all our previous Africa trips. She followed the limo provided to us gratis from Bushtracks in honor of our being dedicated customers, to Ellen and Ron's house, to take photos of all four of us right before leaving. Ron videotaped the arrival of the limo, and the packing of our luggage into the trunk. I could hardly breathe from the excitement. Of course, I was in tears.

The limo doors closed, and we were off on our adventure.

We arrived at SFO five hours before our flight, in order to be first in line to ask for the bulk head seats. And we got them, four in a row. So we were as comfortable as we could be in economy class. Ellen and I helped pass the time at the airport by getting pedicures together. The rest of the travel was a bit harrowing, however:

	Time
Drive to SFO	0.5
Wait at SFO	5
Flight to London	10
Stop-over in London	8
Flight to Lusaka	11
Stop-over in Lusaka	1.5
Flight to Mfuwe	1.5
Drive to Mfuwe Lodge	1
Wait at Mfuwe Lodge	5.5
Drive to Kapamba Camp	4
Total:	47 hours!

We left at 10:30am on Tuesday and arrived at Kapamba Camp at 7:30pm Zambian time on Thursday, with only what little sleep we could get onboard. By the time we arrived at Kapamba I was somewhat zombie-like through our first wonderful dinner under the stars.

But I'm getting ahead of the story. In London we took the Heathrow express and the underground into town, to have our "usual" lunch at Rasa W1. Yummy. We walked by Marble Arch on the way back to the underground, to be able to say that we'd seen at least one tourist attraction while in London.

We were thrilled to arrive at the airport in Lusaka (I was crying, surprise) where we were whisked through customs, had our bags collected and airport visas arranged for us by a Zambian Airways "Meet and Greet" assistant. An uncomfortable snafu occurred in that our flights to Mfuwe had somehow been cancelled - just mine and Jim's; Ellen and Ron still had their seats. Fortunately the mix-up was straightened out and we were able to continue on our way after only a small amount of anxiety. We did miss having our usual breakfast in the airport bar upstairs, however.

Our flight over Zambia was enjoyable and our arrival in Mfuwe confusing and fun, as always, with the various safari camp personnel trying to find their appropriate guests and luggage to load onto their vehicles. Loti introduced himself as the driver for Bushcamps, the company that owns the three camps we were to visit in the South Luangwa. He helped maneuver our luggage below the seats in an open-topped Land Rover for our drive to Kapamba Camp.

We saw our first bird as we were leaving the airport parking lot: a lilac-breasted roller.

The drive to Mfuwe Lodge was filled with fascinating sights, especially the town and all the busy, energetic townsfolk that we passed, riding bicycles or walking, many carrying large loads on their heads or backs: plastic water jugs; sheaves of thatch; bundles of sticks; bags of maize. Many of the little children waved enthusiastically, and the brilliant mix of bright-colored clothing and painted small store fronts set against the red sandy earth and the waving green grasses was magnificent.

As soon as we left the town and crossed the bridge over the Luangwa River into the South Luangwa National Park we began to see wildlife (and I cried again). Ron heard the first hippo calls, Jim saw the first crocodile, and Ellen spotted the first elephant. Our safari had begun!

The short drive from the park entrance to the Mfuwe Lodge was filled with sightings. We enjoyed small herds of zebra. The zebra in this area are called "Crawshays," a subset of the Burchell's zebra, which are noted for their very distinctive black and white patterning, with no "shadow" lines between. We saw troops of yellow baboons with babies, upside-down, clinging to their mother's tummies, and groups of impala and puku. We even saw a herd of over one hundred buffalo. The buffs were mesmerizing to watch … many of them had bright little oxpeckers riding on their backs or heads, looking like clowns, picking at the ticks and other insects. One pair of buffalo mated in front of us, and then the whole herd stampeded across the road before we could pass.

In order that we not have to drive during the hottest part of the day, Bushcamps Company had arranged for us to spend the siesta hours at the Mfuwe Lodge, where we had our first Mosi beers of the trip, a delicious brunch, and enjoyed watching the wildlife in the river bed in front of the expansive deck. We took a refreshing dip in their pool, and were offered one of their chalets to use for a shower before our drive south to our first safari camp. The shower was very pleasant, not only because of its interesting design (sunken tub in a room with open sides overlooking the river), but also because that river was filled with hippos calling quite loudly to each other, right next to the tub! Along the riverbed, in addition to seeing delicate little bushbuck and watching baboon and impala, we also enjoyed the sight of an elephant dusting himself with sand.

At 3:30pm, it was time for our game drive to Kapamba. This had been mentioned to us as being a somewhat harrowing trip, expected to take four hours, and it did take that long. It was quite a lovely drive, however, with captivating sightings along the way. In addition to the four of us and Loti, we also had a "spotter" on board and another couple, Bernice and Neil, also traveling to Kapamba.

We saw magnificent kudu with their striking spiraling horns, scampering warthogs, and our first baobab trees with their deeply crenellated trunks and bizarre-looking extensive network of branches outlined against the sky.

At a brilliant green lagoon we stopped to watch the vibrantly colored jacanas as they dexterously stepped across the hyacinth, as though walking on the water. I'd been keeping track of the bird sightings so far, and we were now up to 18, including hammerkops, herons and plovers, storks and weavers.

Another spot where we pulled over was even more exciting for me: along the muddy banks of a stream were two superb fish eagles standing side-by-side, with their unbelievably white heads and chests, set against the dark brown and black of their wings.

And behind them were two crowned cranes. Glorious birds, very special to Jim and me.

Of course I felt the romantic significance that there were TWO fish eagles and TWO crowned cranes. While we were watching them, deeply moved, each of the birds gave a call. An intensely resonant croak from a crane, followed by the haunting scream of a fish eagle, as he opened his magnificent wings and swooped up to a nearby tree.

Wow.

A sweet little slender mongoose ran by, and we saw our first waterbuck with the disconcerting coloration of a bright white "bull's eye" located precisely on the center of his bum!

The next time we stopped the vehicle to gaze at three majestic male kudu with immense horns, we noticed there were also two female elies behind them, along with a baby. As we watched, one of the elephants suddenly charged at the kudus, and ran them off. It was an intriguing interaction and quite thrilling to see. The elie turned towards us and kept up her pace, just as Loti decided it was a good time to be leaving.

Driving through the mopane forest I waxed romantic again, as these trees have leaves that are shaped like hearts. Preoccupied with these thoughts as I was, I didn't notice the tsetse flies at first but after a few bites on my arms, I realized that they were surrounding us. Luckily the sun set soon thereafter and the flies retired for the night.

We saw our first giraffe in the dusk. The ones that live in this area are called "Thornicroft" giraffe, named for the district commissioner in 1908 who hated to see the giraffe killed simply so their tails could be used as insect switches. He managed to obtain legal protection for them. The lanky animals are quite striking, with very dark distinct borders around their spots. We watched a large group of these splendid animals, including several babies.

As the sky darkened, the Southern Cross appeared, and a brilliant Milky Way wound across the sky.

Our spotter turned on his bright hand-held spotlight and we began our night drive. As always, it was like a tennis match, as each guest followed the light from side to side, scanning the bush. And as we had noticed earlier in the day, Ellen was incredibly good at spotting animals.

Our first nocturnal sighting was a white-tailed mongoose, who posed nicely, staring at us with luminous orange eyes lit up by the spotlight. While we were enjoying the mongoose, Neil suddenly exclaimed excitedly from the back of the vehicle that he could hear that something else was definitely behind us. We abruptly turned around (more to protect our backsides than to see what it was), the spotter shone his light, and we saw a family of four hyena. It was the first time we've ever seen a family of these often solitary animals, and it was great fun to see them walking along together, jostling each other. One of them was very young and fuzzy.

A bush hare hopped away from us, inadvertently disturbing a few elephants by the side of the road. Several of them trumpeted at us, loud as can be, and I was proud of Ellen that she wanted to get even closer, recalling my original fear of trumpeting elies.

And then we arrived at our first safari camp, situated on the bank of the Kapamba River near its confluence with the Luangwa. The camp was lit only with paraffin lanterns, scattered all about the grounds, their twinkling radiance warm and inviting. I cried! We were provided warm facecloths scented with lavender, and Ollie, the camp host, showed us to our chalets.

Although hard to discern in the dark, our room was fantastic, with lots of space, smooth concrete floors and a king-sized bed with mosquito netting, surrounded by stucco walls on three sides. These walls are only 6 feet high, and the thatch roof is set higher so that there is an opening above the walls of several feet, looking directly out into the bush. The fourth wall of the building is completely open, overlooking the river, although at night we pulled an interestingly designed set of iron grillwork (which looked like spider webs) across this open space, to protect from leopards jumping in and joining us in bed. The attached loo, also with a completely open side facing the river, is sumptuous. The sunken concrete tub could be filled within five minutes with steaming hot water from the boiler. Double shower heads, double concrete sinks, and deliciously scented shampoos and shower gels complemented this lovely bathroom.

When ready, we were escorted back to the sitenge. This open thatched building had a bar area, a large dining table, a set of couches for reclining, and a large deck overlooking the river. On this first night we had a barbeque out on the actual riverbank, so we walked out onto the sand where the dining table had been set, complete with white tablecloth and candles, china and stemware. We had a pre-dinner Amarula, our first Amarula of the trip. The dinner was delicious, as always, made from fresh ingredients most of which were grown locally.

Thai lemon-grass/coconut soup
Fresh-baked rolls
Rice/veggies/meat/chilies wokked to order over an open fire
Banana/toffee/rum also wokked over the open fire

I decided this year to keep track of our daily menus, to demonstrate the decadent fare to which we were always treated. Breakfast was the same at each camp: tea or coffee, porridge heated in iron pots over an open fire, freshly baked bread toasted over the open fire, freshly baked muffins, fresh homemade jams and marmalades, and fresh fruit. Morning tea, during our game drives or walks, included a freshly baked snack or small cookies along with tea/coffee, or cool drinks if the day was warm. Afternoon tea, provided after our mid-day siesta and before starting our afternoon game drives or walks, also included freshly baked cakes or cookies, along with tea/coffee or cool drinks. All breads were made fresh daily at each camp, baked in underground "ovens." Sundowners always included Amarula for Jim and me. Drinks were offered before dinner and wine liberally poured throughout dinner. Port or Amarula was offered after dinner, and we usually had a Mosi beer with our brunch, when returning from our morning drive or walk.

During the night, I slept quite soundly with only a tiny bit of jet-lag in the early morning. I heard one lone hyena moan, and buffalo calling quite nearby. And myriads of frogs croaked loudly! We always kept one paraffin lantern lit during the night in the loo, so when getting up at night we would not have trouble finding our way.

We were awakened at 5:45am with a cheery "Good morning!" and dressed while listening to the deafening dawn chorus of birds. After a quick breakfast we began our first activity with our guide, Dean.

As an aside, each of the three "Bushcamp Company" properties where we stayed in the South Luangwa on this visit was operated in the same fashion: each camp has only four chalets, overseen by one host, one guide, and eight staff. The staff included the cook, assistant cook, food server, barkeeper, house keeper, laundry person who washed and pressed daily, groundskeeper, and a scout required for walking safaris in the National Park. Daily activities were the same as at all our previous safari camps, however, in these camps there is only one guide and one vehicle so the guests would decide each morning and afternoon on either a game drive or a walking safari. This design allowed for the guests, staff, and guides to become quite close-knit and we really appreciated that intimacy. The staff are endearing and hard working; the hosts are charming and helpful and fun; the guides are super.

Our first early morning drive was slow and quiet, and we truly began to feel part of the African bush. We sat watching a single adolescent puku, appreciating his lush fur and skittish behavior, against a background of sparkling spider webs amongst the tall elephant grasses in the morning dew.

We saw many bird species, learning all about their calls and behavior. We had now seen 57 species on the trip so far, including one life bird: a little yellow white-eye. We got out of the vehicle to get a close-up view of the striking "scrambled-egg tree," with its arresting yellow flower globules against dark brown trunk and branches. Towering ebony trees were laced with vines covered in bright red flowers, "flaming combretum." We pulled the seeds off stems of wild lavender and basil, deeply inhaling the luxurious scents. We stopped at a muddy water hole and listened to the Heuglin's robin warbling effervescently. Ron spotted the beautiful bird first, with its deep russet chest and intense black and white "zebra" head. Dean picked up handfuls of sausage-tree flowers so we could enjoy their smooth large red petals and intoxicating scent, before placing them back on the ground for the impala who love to munch them.

It was such a peaceful morning! We watched a group of three giraffe for a long time, learning to differentiate between male and female. The male is typically larger, with a bumpier head, and the hair around his horns was worn smooth from sparring over territory. He was quite interested in one of the females and followed her closely, even nudging her at one point. She was playing hard to get, however, until he suddenly bent down and gently rubbed his face all the way up along her back to her neck. They moved off into the bushes together, and we all hoped he got lucky after that sensuous move!

Dean spied a group of elephants on the far side of the Luangwa River (or was it Ellen who spotted them first?), and suggested that we trek over to get a closer look. We walked across the large sandy stretch of riverbank in our boots and when we got to the water's edge we took them off, splashing through the river in bare feet. The water was only a few feet deep, nonetheless it was exciting to be in the river hunting elies! On the far side we clambered up the rather steep bank and stood in the tall grasses with a view of elies between the stalks, really close. They didn't seem to mind us at first, but soon realized our presence (the wind was not in our favor) and slowly moved away, with the characteristic creakings and snappings of elephants moving through the bush.

Back at camp, we had our first outdoor showers, glorious warm cascades of water as we stood in an elegant smooth concrete tub looking directly at waving grasses and the Kapamba River. Brunch was served, and then it was time for a rest. During this siesta we actually took naps, although in future days we spent more time writing in our journals, taking bubble baths, or quietly bird watching around the camps.

Herb bread
Spinach/leek fritters
Spare ribs (moussaka for Bernice and me)
Couscous with veggies
Green salad
Freshly cut pineapple

We went for a walking safari in the afternoon, and Dean had suggested that we only wear flip-flops as we were going to start the safari in the river. It was pretty cool to be walking along in the river; accompanied by our scout - I've never actually seen a BAREFOOT man slinging a RIFLE! We traversed the river, enjoying baboons along the shore, then donned our flip-flops to hike into the mopane forest and tall grasses on the far side. We visited a "forbidden pool," forbidden since it often has a croc or hippo in it, and wondered how this waterhole could stay filled with water even during the dry season, in the eerie twilight under the thick leaf canopy. We collected berries from ferettia bushes that can be used as an aphrodisiac and discussed that the orchids in the ebony trees are epiphytes, not parasites. Walking through the deep grasses and mopane trees was work but Dean postulated that one needed to undergo hardship to appreciate the ease and grandeur of traveling along the open river way. For me, I was secretly happy that the bushes were scratching my itchy tsetse bites.

When we finally re-emerged to the river we enjoyed the sight of an entire troop of baboons leaping and cavorting across the water. There was a stick in the middle of the stream, sticking up above the water a few feet, and the baby baboons would climb the stick and leap off of the top, rather than gallop around it.

We walked up the river, each couple hand in hand, just as the sun was setting over the Machinga Scarp Hills behind, and stopped for sunset photos along the sandy bank. We discovered a set of elie footprints, complete with splashed water, leading away from the river and realized that the group we could see in the bushes was the same group we'd seen earlier, but now on the other side of the river. There was a baby in the group - I sighed, very contented.

Around a bend in the river, we came across a charming sight: deck chairs sitting in the river, with lanterns hung on tree branches set in the sand, a table with our sundowner drinks, and a little brazier (also in the river) heating nuts for our snacks. As we sat with our feet in the cool shallow water, toasting each other with Amarula and munching warm nuts, I cried to see the tiny orange smile of a new moon peeking out at us where the sun had set.

After this relaxing moment, we were trucked back to camp to get our night drive kit. For the next hour we zipped around, driving through tall elephant grass in the dark, looking for nocturnal life in the bush. We saw our first elephant shrew with its adorable long snout and brilliant shiny eyes, another bush hare, a Sharp's grysbok, and a genet.

Back at camp we made a quick trip to our chalet to drop off our gear, and Jim and I sat on the porch, holding hands, enthralled, gazing at the tiny moon sliver while listening to a bull elie munching and crashing and gurgling and snorting in the grasses near by.

Tomato/goat cheese terrine
Rolls
Bream (tilapia) crusted with nuts
Scalloped potatoes
Lemon tart

During dinner out under the stars, I kept a cold rag on my leg to try to help with the swelling from the tsetse bites. I counted 10 on my leg, and another few on my arm, and one on my forehead, all of which had swelled up uncomfortably; my skin was stretched taut and hot to the touch. I tried ibuprofen, Sudafed, benadryl, topical cortisone, and topical anti-histamines but nothing seemed to have any effect. Jim, for his part, had a bout of tummy trouble during the night which luckily was gone by morning. We weren't feeling particularly much like intrepid explorers!

In the night we heard a lone hyena again, a mysterious lilting "whooo-oop!," and the loud cough-like sound of a leopard in camp. We'd never heard that before. Very exciting.

The next morning we had a long drive through a mopane forest, unfortunately infested with tsetses. I began referring – affectionately - to this woodland as "the valley of the shadow of the evil tsetse," or "the dreaded mopane forest of death." Ron suggested we call it, "Cynthia's delight." The dense thickets were simply swarming with the creatures, and I eventually wrapped up in a wool blanket to try to ward off the number of new bites.

The drive began along the river, where we enjoyed the flitting of fork-tailed drongos - I thought of Kalpana. While everyone else was enjoying two stately giraffe, I was certain that I saw a pied avocet in the river, an unusual sighting and therefore possibly suspect - I was convinced based on its upturned beak and striped sides. The giraffe began staring intently at the other side of the river and when we heard a puku alarm call in that same direction, we drove quickly across the river along a line of submerged sand bags, hoping to see what was going on over there. We never did find a predator and the alarm call subsided. We did, however, learn that hippos eat about 40 kg of vegetable matter per night, while giraffe eat 30 kg, and female giraffe return to the actual site of their own birth to have their babies.

We collected river-washed rocks from a hillside – it was interesting to note how they were located up so high on the hill, far above the river, indicative of how things were different in the past. On our walk around the hilltop we saw a duiker sleeping nest in a stand of "love grass," a latex rubber tree, a pepper-leaved camiflora from which myrrh is derived. It can be burned to ward off evil spirits, or used as a natural antibiotic, and its bark performs photosynthesis rather than the leaves. We also heard a honey guide calling us, and Ellen and I spotted an Arnot's chat, another lifer for me.

Back down by the river we watched a giraffe drinking with his legs spread way apart, splashing the water all around his head as he sucked it up from the river. We sat through a few "twisters," dust devils made of leaves that floated all around us. A puku sprinted up the river as majestic vultures hopped along the sand, leaping to take off. Several vultures circling overhead were being dive-bombed by a martial eagle showing that he is a much better pilot than they.

Balsamic onion bread
Lasagna (vegetarian servings baked in individual ramekins)
Potato pizza
Greek salad
Spicy 3-bean salad
Tomato/avocado salad
Fresh fruit

After our delicious brunch, I took photos of the staff and Jim printed copies for everyone as gifts. I offered to each staff member that he could have his own solo photo taken doing whatever he wanted, and each guy chose to pose doing his job. They were so honored by this opportunity to be able to show their kids back home what they do, and were so proud of their work. I also took a group photo, everyone jostling and giggling and teasing each other, and gave them each a copy of that scene. It was particularly fun to watch the staff watching the printer in action; they were also quite impressed with the solar panel kit that Jim built.

As a "thank you" for the photos, the guys procured for me a fresh sausage-tree fruit, and said that the juice should help with my tsetse bites. I was willing to try anything at this stage, so I borrowed Jim's knife, cut a fresh slice, and rubbed it on the swollen skin. AMAZING. Within hours, I had relief. My leg hurt less and I was able to walk without pain. At night I had been sleeping with the leg elevated on two pillows, cooling it with a wet washcloth. Even though the sausage fruit turned my skin an unusual shade of orange, I was overjoyed. It was such an improvement that I even noticed the upper section of my arm, which had been swollen taut, was now all loose and flabby again, for better or worse! ☺

We had another wonderful afternoon game drive. We learned about the "ordeal tree," which has poisonous leaves that were used for determining the guilt or innocence of putative criminals - a trial by ordeal. If a person lived after being forced to drink the toxic leaf concoction, he was obviously innocent. The accused criminal hoped that a friend was the person making the infusion, as a strong brew would induce purging and could be survived, whereas a weak mix was lethal.

We watched a superb bull elie by the riverbank, carefully making his way along the sandy cliff and removing the grass as he went, twirling his trunk around a bundle of stalks and yanking the grass out by the roots. He'd then shake the bunches to get rid of the sand, even going so far as to thwack his face with the bunches to dislodge every bit of dirt before shoving the grass into his mouth. We listened to and gazed at many birds, and witnessed pronking puku, which is quite a sight. We stopped to enjoy a dazzling brown-hooded kingfisher perched on a tree branch, only to realize that the trees above were filled with vervet monkeys, leaping from bough to bough. We enjoyed sundown – and our sundowners - by a large pod of hippo in the Luangwa River, watching gigantic bulls giving amazing yawn displays, and enjoyed tiny hippos practicing with their little yawns showing off their pink mouths with no teeth. All the while, we listened to a constant explosive burbling and guffawing. In the distance we could even discern the sounds of a serious hippo battle for territory.

The brilliant orange moon-smile appeared in the sky, along with Jupiter and the rest of the starry congregation. Our night drive back to camp was uneventful, with one bush hare being the only nocturnal animal sighted. Nevertheless it was transcendent to drive through the bush in the dark, listening to elies munching in the tall thickets.

At dinner we celebrated with a bottle of Opus One 2001 that Ron had brought all the way from home, very carefully wrapped in his luggage! It was delicious and terrifically romantic to be sharing this special wine in such an exotic setting.

Gem squash filled with corn and cheese, baked in the shell
Rolls
Chicken (or cabbage) with glazed apples
Rice
Snow Peas
Chocolate soufflé in individual ramekins

And then to sleep, lulled by croaking frogs.

In the morning we went for a walking safari, starting from camp and wandering along the riverbank. We strode quite close to several male buffalo in the tall grass, only to hear them galloping away. We then identified francolins whining to locate each other in the tall grass, all the while teasing Ellen for mistakenly calling them "Lincolns" (which eventually morphed to "Madisons" as the trip progressed). They also have the nick-name of "heart-attack birds," as when accidentally flushed they make an incredible racket that would startle anyone. We enjoyed several butterflies, a dove taking a bath in the river, and identified porcupine poop. While we were resting on a log in a sandy cove, we learned from our guide that when walking on sand, one should step IN the footprints of the person in front, as the sand is compressed there and the trip is made much easier. We saw lion and genet tracks and compared the difference in size, noticing it is actually quite significant. On a muddy bank we spotted a vast number of tiny baby frogs, slightly smaller than M&M candies, hopping around. Jim wanted to hold one and succeeded, if only briefly!

Before too long we came to a charming lagoon, mostly dry, where we sat in the soft grass-covered dried mud elephant foot-holes, and consumed our tea and almond cookies. We flushed about twenty guinea fowl, fat little biddies who spiraled up to the sky with a silent flapping of wings, and then watched several swallow-tailed bee-eaters - a life bird for us - genuinely stunning birds with bright turquoise tails and gleaming green tummies. They flitted for insects, fought with white-fronted bee-eaters, and perched on grass stems.

As we trekked back to camp through a burned-out mopane forest (yes, there were more tsetse flies), I entertained myself by composing a song:

> Walking along in Zambia,
>   Silent, single file;
> Following leopard's spoor
>   We rack up another mile;
> Dreams of brunch and dinner,
>   Complete with fresh-caught bream;
> We're the Africa 2006 Assault Team.
>
> Elephants chasing kudu,
>   Hippos yawning wide;
> Bouncing along on night drives,
>   Searching from side to side;
> Bush hares and genets scamper,
>   White-tailed mongoose eyes gleam;
> We're the Africa 2006 Assault Team.

Cynthia's covered in tsetse bites,
    Ellen spots all the game;
Ron brings us Opus One,
    Jim prints photos for staff to frame;
Sneaking up on elephants,
    Leaping impala are like a dream;
We're the Africa 2006 Assault Team.

Our villas look out on the river,
    We see baboons, buffalos, and birds;
Paraffin lanterns at night,
    While puku gather in herds;
Soaking in a sumptuous bath,
    Drinking coffee with cream;
We're the Africa 2006 Assault Team.

I sang the song during brunch, and tried hard not to cry, as we were soon to leave for our next camp. We took photos of the two staff members who hadn't been able to have them taken the day before. One of them had insisted on waiting until his camp shirt was dry, as he'd just washed it that morning, before he would be willing to be in a photo! It is very sweet to see how serious and appreciative these guys are about the responsibility of their jobs. I also gave the four T-shirts that we'd brought as gifts, to Ollie to distribute to the staff. He later explained to me that he needed five T-shirts to please everyone, although only if I had another one to spare. I went back to our chalet and brought him my Mount Whitney T-shirt (which I'd been planning to sleep in at night …). It was fun to tell Ollie the story of each T-shirt, knowing how they will be cherished.

Corn bread
Stir-fried zucchini, cauliflower, peppers, broccoli
Breaded bream
Potato wedges
Beet salad
Green salad
Fresh papaya

And then it was time to say goodbye to Kapamba Camp, and a rather emotional time. We'd become so attached to everyone over the past few days, Ollie and all the guys. It was hard to leave, but what a wonderful experience we'd had.

Dean drove us to our next camp, Kuyenda, about a two hour drive through the heat of the day. Ollie had provided me with my own personal spray can of "DOOM," which I could use to directly kill the tsetse flies as we were going to be driving through another dreaded mopane forest. Luckily, I didn't have to use it too often as it was probably quite unhealthy to inhale such strong insecticide.

Getting playful in the jeep, Jim mentioned that since this was officially siesta time, we ought to be able to "go at it" in the back seat and Ellen proclaimed it to be a great idea, since then we could use the bumps of the road "do all the work" for us. We joked about the "Land Rover Club," similar to the more famous "Mile-high Club," and imitated Jim's calling out: "Dean, speed up! ... Great, ok ... Now, turn left! ..."

Dean's rejoinder to all this was, "she'll have to hold on," to which Jim smugly replied, "She has to anyway."

We thought we were quite hilarious. It reminded me of all the jokes we'd recently developed, regarding my bringing my sausage fruit with me wherever we went. It had become one of my most prized possessions as I slathered the juice on my bites whenever I had the opportunity.

On the drive we stopped at several exquisite lagoons and enjoyed astonishing wildlife; waterbuck, baboons, impala, herons, geese, warthogs, as well as four majestic male kudu staring at us from a wooded glade.

Dean had told us that the camp manager and guide at Kuyenda were probably the most famous in all of the South Luangwa, and we joked about how we were going to tell them all about what we had learned from Dean, so they'd gain an appreciation of his skills. We intended to get the names of the trees and birds all wrong, and then say that that's what Dean had taught us. Ellen joked that we should only whisper to them, even when saying "hello" when first meeting them, and then explain that Dean insisted we could only whisper in the bush. Again, we thought we were insanely funny.

We came across two elies, an older and a younger female, hanging out together. A little later, we came across another two elies, and Dean pointed out that these were, "the same configuration as before."

"What, two big ears, four legs, and a trunk?" Jim queried.

More raucous laughter.

We finally arrived at Kuyenda Camp, situated on the dry river bed of the Manzi River. This was an authentic traditional bush camp, with no electricity and the chalets were simple round thatched huts made from elephant grass, sorghum stalks, and bamboo sticks. They required five weeks to build, the camp was open for five months, and then it took five days to dismantle them in advance of each rainy season. The stalks came from villagers harvesting sorghum; normally they would simply be thrown away but in this case the villagers could collect money for them. Termites love to eat the sorghum, and we could hear the pleasant ticking sound of their eating away at the inside of our room's walls.

These round rooms had concrete floors covered with reed mats, but the attached bathrooms had no floors or roofs ... just sand and sky! The front section of the hut was open but chicken wire netting was provided to discourage nocturnal animals from joining us at night. As at Kapamba, there was a large gap above the walls and below the thatch roof as well, so that the rooms felt quite open. The beds had mosquito netting draped around them at night which was particularly romantic. I loved the outdoor loo; the shower was simply a 50-gallon drum on a wooden tower beside the wall; when one wanted a hot shower the staff would add some heated water to the drum. A small slate floor below the shower was covered with bamboo mats to stand on while drying off. The double sink bowls were set into a large log.

The main sitting room was similarly simple, with a thatched roof and concrete floor but no walls, offering wonderful openness to the bush.

The hostess, Babette, and the guide, Phil Barry, are legendary, so we were thrilled to have the opportunity to stay with them in their camp. We started with a cold Mosi beer and met the other guests: Mike Brearley (apparently a quite famous ex-cricket captain from the UK but of course none of us Americans knew of him) and his sweet East Indian partner Mana and their daughter Lara, who currently works at an NGO in Zambia.

We were asked to select which hut each couple wanted to stay in, as one of the huts had a king-sized bed and the other had two queen-sized beds. We both thought the bigger bed was a nicer choice, so to be fair we flipped a coin. We weren't able to find a coin amongst us, though, as we'd not yet had to use any money on the trip. Fortunately, one of the British guests furnished a coin and Jim and I won the toss (Scottish flag versus The Queen!). The staff took off with our luggage to each of our respective huts.

After our beer and snack, Phil took us on a short walking safari of the local area around the camp. It was spectacular, with several small pools, mineral springs and marshes nearby. Simply myriads of birds live in these wetlands. I loved seeing the Madagascar squacco heron, a very rare migrant who comes to Phil's lagoon every year (of course, a lifer for us), along with a painted snipe, black crake, and a yellow-throated longclaw. Dazzling colors!

We also had a lot of dung discussions, learning that the palm nuts that can be collected from elie dung are typically found only in the dung of males, as they are large enough to shake such a sizeable tree as a palm. And the palm nuts can be shelled to reveal the pretty white "vegetable ivory" pod inside, such as the one we collected last year that was carved into a hedgehog by Jessica for Jim's birthday present.

After sundowners we followed up with a night drive, and really enjoyed the little bush hare right outside of camp who simply wouldn't leave the road. He just kept hopping along in front of the vehicle, inside the track, even when Phil revved the engine and threatened to flatten him. We also saw a great African hawk eagle in a tree, looking like an owl, three genets, and an elephant shrew.

Dinner around the beautifully-decorated central table was a fun affair, filled with interesting discussions and safari stories.

Spinach quiche-ettes
Rolls
Sausages
Bream or stuffed butternut squash
Glazed carrots with homemade mango chutney
Snow peas with almonds
Parmesan potato wedges
Apple crisp with cream

I slept soundly, although a Scops owl in a nearby tree called ALL night long, every four seconds on the dot, as I can attest from my various trips to the loo and slicing of sausage fruit for my few new tsetse bites during the middle of the night.

Our "good morning" was accompanied by a pitcher of hot water, pleasant for face washing while standing in the cool sand in the bathroom. Jim found a toad in his boot (luckily before putting his foot in it), and then shook a lizard out of his jacket sleeve. Camp is the complete wildlife experience …

We met Ellen and Ron at the sitting room for breakfast, and Ellen was not looking very happy. "Have you slept well?" I asked her, the proper morning greeting in Zambia, and her response was somewhat strained.

"I am worried about the bees," she replied, explaining that their hut was right under an expansive sausage fruit tree in full flower, and that the bees were not only up in the tree but also ALL over the bathroom floor, sipping nectar from the flowers that had dropped on the sand overnight. As she has a terrible reaction to bee stings, she had been uncomfortably nervous to use the loo. Not the way we're supposed to feel while on vacation. Jim and I happily agreed to switch huts particularly as the choice had been decided by a coin toss, anyway. It turned out that we found our new hut to be even nicer for our purposes anyway, as Jim had so much equipment that he was able to use the second bed as a staging area for all our accoutrement.

After a hearty breakfast of porridge and toast, Phil turned from the dining hut towards the staff quarters and loudly called: "I … zay … ahhh!" in a sing-song voice, and was answered by a melodic loud chorus of sung "whooo … hoooo …!" from all the staff. This had become quite a tradition, informing our scout Isaiah that his services were needed.

We had an extraordinary walking safari, appreciating Phil's infectious respect and interest for everything in the bush. He spoke of how many animals the sausage tree fruits and flowers fed, including zebra, porcupine, hippo, giraffe, squirrels, and impala. The strikingly beautiful winter thorn tree (so named because it is the only tree that is NOT leafy and green in the wet season), is also important food for these animals. We enjoyed a lovely lagoon covered with hyacinth and very red algae, and watched birds and giraffe drinking. The giraffe hugged their necks around each other at one point. Jim commented that it was a head-butt, not a hug, but I chose to leave that open to interpretation. And the ubiquitous white-crowned plovers frantically called and called at us, warning us away from their eggs buried in the sand.

Jim collected handfuls of lavender seed pods into a plastic baggie, to bring back home for use in making scented soaps.

When we reached the bank of the Luangwa River, we enjoyed baboons playing amongst tree trunks on the sandy cliff, and hippos grunting and harrumphing in the water. We learned all about crocodiles as we observed some huge specimens basking in the sun. Like turtles, the temperature of the sand in which the eggs are incubated determines the sex of the infant, and the croc mothers are apparently quite attentive to these eggs; protecting them from predators and helping the babies to dig themselves out once they leave the eggs. If there is danger present when the babies emerge, the mother even carries them delicately in her mouth to a safer site to begin life. Until now I had simply thought of crocs as "evil incarnate," so these tender behaviors made me hate them a little less.

We tasted ebony fruit, small hard round berries that were slightly sweet. I found the snake and/or mongoose holes bored into an extinct termite mound quite fascinating, and thrilled to the sight of ground hornbills flying. They are rather ugly creatures when they strut along the ground, all black with a gargoyle-like red and black head, yet they become elegant as they lift off, showing their brilliant unexpected white patches on their giant wings. They are interesting to watch as they throw their food – small insects and lizards – up into the air first, and then catch them in their bills for eating.

And then, we came across elies! There were several females drinking from a beautiful bright green pool, and although the matriarch sniffed towards us frequently, twining her trunk around in our direction repeatedly and questioningly, they didn't move on for quite some time. There was a young male who crossed our path and seemed intent on joining the females, and Phil explained that sometimes the young males don't quite "get it," that they are supposed to leave the breeding herd at a certain age and strike out on their own, and so they hang around their moms and sisters for a while after first leaving. We also came upon another young male elephant with a broken ear: the cartilage on the right side was defective so that he couldn't keep the ear from blocking his vision. Phil wanted to make sure he knew we were there, rather than suddenly startling him as we were on his "bad" side, and so he coughed loudly and tapped his walking stick on the ground. We waited until the elie had turned his head towards us and actually made eye contact, before continuing on our walk. An informed elephant is a safer elephant!

And, as we climbed out of a gully and reached the road, there was Babette, arriving with a jeep to take us back to camp. Impeccable timing! Before we left for brunch, Phil showed us a truly alien-looking strangler fig, growing around an ebony tree. The strangler grows from a seed deposited on top of the tree, in feces from a bird, and then the roots grow down to the ground and it eventually kills the tree around which it grew.

Beef/egg casserole
Chicken empanadas
Lentil/rice salad
Stir-fried veggie salad
Butternut squash "coleslaw" with sunflower seeds
Green salad with feta
Tomato/avocado salad
Fresh fruit

After our extraordinary brunch and refreshingly wonderful siesta with a lingering hot shower, and luscious pumpkin/date bread for tea, we had an afternoon game drive. A herd of zebra posed for us, and a baboon mother with a tiny baby literally barked at us as we passed.

Phil drove on his detour around the sausage tree next to the road which earned a giggle from his passengers. Dean had told us the story of the time that a guide had been bonked in the head with a sausage falling from a tree, and subsequently Phil had engineered a by-pass AROUND this tree, which he always took rather than just quickly driving under the tree and taking a chance. Some people might think a quick drive-by was not that much of a risk but as we were laughing about the detour, Phil pointed out a huge new sausage which had fallen right on the road subsequent to our morning drive! Phil got out to pick it up and show it to us - it was quite impressive and must have been about three feet long, heavy as a bowling ball. Although we made a few jokes about the size of Phil's sausage, we never laughed about his taking the detour after that.

We had wonderful elie experiences! One young male seemed annoyed at us, as he impatiently swayed his foot, pawing forwards and backwards at the ground, while staring at us pointedly. He then grabbed a heavy tree branch with his trunk and maneuvered it into a better spot for munching the bark. We saw a few more elies down in a ravine, digging in the sand with their trunks and sucking fresh water from the deep holes. It was interesting that they would go to this much trouble, rather than easily drinking from the pond which was only a few feet away. Was the water in the lagoon not clean enough? One of these elephants also appeared annoyed at us, and gave a deep rumble and a head shake as we passed.

A baby puku paused to stare, standing next to both mom and dad, before all three leapt away in stunning formation. A dozen banded mongooses ran along a grassy bank, looking like a tribe of rats with furry tails.

And then, we saw it.

Our first LION!

He was asleep under a bush in a very dense thicket, and had it not been for the safari vehicle stopped next to the bush with all inhabitants peering over the edge in excitement, we may have simply driven past without seeing him. After a few moments, the other vehicle left and we got to observe the lion in solitude. He wasn't very active, simply yawning a bit now and then, and licking his paws, quite LARGE paws, I might caution, and purring deeply. After we'd watched him for some time, he decided to get up and stretch so we could see his stunning face and generous blond mane.

Phil thought there were probably other lions nearby as he heard some additional growls, and so we drove around the thicket and came to the other side. There, lying in the low grass, was a breeding pair! They were both sleeping (who wouldn't, after mating every fifteen minutes for several days!), the female sprawled on her back with legs akimbo, the male in a more prim reclining position with his head held up but lolling as he closed his eyes. This male had a luxurious, thick dark mane. Again, we watched the cats sleeping for quite some time before there was any movement of significance.

And then suddenly, Mr. Lion stood up, stretched, and slowly sauntered around behind Mrs. Lion, who conveniently rolled onto her tummy … he maneuvered his way forward till he was standing astride her backside … and within seconds they had mated, right in front of us! The act was so quick we almost missed it, but the incredible part was seeing the expressions on their faces, contorted with effort, curling lips and wide-opened jaws, snarling and biting, as they suddenly built to the climax.

After this spectacle, they both flopped back down on the ground, spent. Just then I heard the characteristic branch-breaking and twig-snapping from a line of trees across the marshy meadow, and a group of elephants appeared.

We watched as they majestically strode to the mud in the middle of the open field, and began searching for foodstuffs in the bog. The obvious matriarch and another large female came closest to our vehicle, while a tiny baby was in the middle of the field and there were more females behind. The baby was so small he still had a very fuzzy back, and his ears were laid flat against his head. Tremendously adorable!

Phil watched the large females closely, as they were clearly perturbed by our presence. The matriarch swayed her foot, rumbled quite loudly, and then blustered repeatedly with her ears flopping madly while exaggeratedly sniffing in our direction. Phil suggested we move on after a bit, so as not to irritate them further.

We came across four baby francolins (I don't like to use the newer term for these birds, "spurfowl"), taking sand baths in the middle of the road. The mother was off to the side, searching for food in the grass. They would flop down and fluff up, swooshing dirt all throughout their feathers, then leap up and run along the road, skipping like little kids, then flop down and do it all over again.

Watching this performance while the glorious sun was setting over the trees, I finally had a good long cry. The lions, the elies, the baby birds, the sunset … I LOVE it here. The quiet and the scents and the FEELING of Africa.

Sundowners were taken out on an open plain, with white tablecloth on the tailgate and an elegant bar setup, snacks on skewers, listening to repeated long, low, lion calls – punctuated with a crowned crane plaintively crying – and a brilliant orange slice of moon and Jupiter sparkling above.

Our night drive commenced and we saw the usual genet and white-tailed mongoose, and heard a Scops owl.

Before we barely had time to consider what it was, we saw our first leopard!

He was in a deep thicket, although fairly visible from the road, sleeping with his head down. We were able to admire his stunning, luxurious, healthy coat of fur, covered with a fascinating pattern of black rosettes with orange centers. After we watched his recumbent form for a while, we each began to wish that he'd get up and DO something, so that at least we could see his face. I could hear Phil whispering under his breath "Come on, get up! Come on, get up! Come on! Get up!" I was touched by how much he clearly cared.

Finally it seemed that the leopard listened to Phil, and gracefully stood up, looking straight at us long enough that we could admire his feline gaze and his remarkably thick neck. What a handsome face! But then, he was on the move, choreographing a way through the thick shrubs with surprising speed. We tried as best we could to follow with our lumbering jeep, driving over anything in our way, including small trees, but he soon slipped away into an impenetrable thicket.

Phil sat quietly thinking for a while, deciding where we would be able to intersect the cat's peregrinations. "He thinks like a leopard!" Babette explained later.

We drove for quite some distance, in what seemed like the opposite direction to me when suddenly, there was the leopard again! Very impressive. This time we were able to follow him for a longer time as he was walking parallel to the road on which we were driving, but all the time he was nevertheless behind a screen of bushes and tree trunks, and even with the spotlight shining on him it was difficult to get a perfect view. Finally, he angled off into the distance and we let him go about his business in the night.

We passed a few elies on the way home, and the spotter was very good about not shining the light directly in their eyes. We did have to "encourage" one bull elephant to walk OFF the road with Phil's polite but emphatic tapping on the metal of the jeep, shining the light on the elie if he turned towards us, and removing it when he moved away from us. This maneuver worked, and he eventually trundled off into the woods. The next elie we drove past was close to the road, on Ron's and my side of the vehicle. It was quite the adrenalin rush for the two of us when she turned, mock charged, and loudly trumpeted right at us.

Back home, lanterns set all around the camp gave off a warm comforting glow. As usual, dinner was a lavish affair with an elegantly laid table, sausage fruit flowers at each place setting, and simple white linen napkins folded like cranes.

Pumpkin-apple soup
Rolls
Mashed potatoes
Broccoli/cauliflower
String beans
Stir-fried veggies (zucchini, peppers)
Beef stroganoff
Crepes Suzette

We had a fascinating discussion over dinner of Phil's astonishing animal conservation experiences over the years, his not only managing to save countless snared animals but even rescuing antelopes and elephants who had become mired in mud. I imagined how difficult that would be, hours of back-breaking work, and I was happy/honored to be spending our time with this dedicated individual. Later, I was sent to sleep by the joyous sound of an elie chomping by our tent during the night.

Our morning walking safari was wonderful again, enjoying secluded lagoons covered with intensely green hyacinth, trees smeared in mud where elephants had scratched themselves by rubbing against them after wallowing, hippo dung scattered over low bushes for territory marking, fresh leopard dung, and aardvark holes. All the while, we listened to the Cape turtle and red-eye doves incessantly calling and the white-browed sparrow weavers chatting wildly. Just another walk in the African bush.

We saw three large elephants ahead with a baby, and Phil scuffed at the ground with his foot. When they ambled off he mentioned that they must have heard us, as the wind was in our favor. We viewed an excavated cliff area which looked as if it had been cleared by road work vehicles. However, Phil explained that it had all been done by elephants digging for minerals. Other animals like impala and buffalo would come and lick for the salts too. We enjoyed termite trails and mounds all around – again it seems as though it's all about dung and termites! We looked closely at a warthog skull to see just how their tusks rub against each other to become razor sharp.

We passed an ordeal tree again and learned something new about this fascinating plant: there is only one kind of caterpillar which can eat its poisonous leaves, able to somehow metabolize the toxin into a harmless substance, as many birds eat the caterpillars. As an aside, we also learned that civets are the only animal that can eat poisonous millipedes. Someone is always someone else's food and there's no waste out here.

A herd of zebra allowed us to get quite close, and while having a break for water and snacks under a tamarind tree we tasted the pods, which were as sour as lemons … on steroids! We saved mahogany seed pods and pod cases, and vegetable ivory which had already been removed from the hard palm nut by virtue of baboons. More poop went under examination, as we saw a midden of civets, who share a communal loo yet make really large individual samples, and an impala midden where all the males go on top of one another's droppings, thereby trying to assert dominance. It was quite noticeable that the grass was always greener around these collections due to the nitrogen content.

We also observed that elephant footprints in mud are really huge; they are able to get out of mud more easily than most animals, as their feet expand when they step down and so when lifted, the foot becomes smaller in diameter. So interesting! We saw fresh growth on a termite mound, meaning that rain may be coming soon - how DO they know these things?

We watched elies climbing the bank of the Luangwa River, and cherished watching a large female helping a baby by pushing it from behind. We had elephants on all sides at one point of our walk, and were threading our way carefully along when at one point we backed up and altered course due to an elephant IN our path. "Shall we leave? What a good idea!" laughed Jim and I to each other.

Bream cakes
Baked rolls, with spinach, bacon, egg, and cheese filling
Eggplant/pepper casserole
Pasta with fresh veggies
Boerewors (beef sausage)
Fresh beets
Green salad
Fresh fruit

After lunch we took a siesta, with hot showers and some nice down time to relax. Jim and Ron tried to figure out how to fix Babette's satellite connection, while I watched birds in the trees and around the small birdbath. Very peaceful.

In the afternoon we went for a game drive, and in total, saw 18 large animal species! We saw zebra, baboons, and impala hanging out along the banks of the lagoon, in the perfect afternoon light. We watched a nursing puku, which was adorable but also head butting her mom with incredible vigor. A band of banded mongeese ran by in a horde, and then we began our elie encounters.

First we saw about eight of them in a thicket, with a baby hidden in amongst the large females. The matriarch, right next to our vehicle, flapped her ears at us loudly … thwack, thwack, thwack … and sniffed at us repeatedly, posturing and blustering. We moved on and came across another family group, of about ten individuals, crossing the road ahead of us, one by one, with lots of sniffing and ear-waving and deep sonorous rumbling. Another four elephants came by next, and then another two. I was in heaven!

We took photos at sunset and drank our Amarula with wonderful cheese biscuits which I decided were shaped like elephants (upon closer inspection I realized they were rhinos), and Ellen found some potshards – we were clearly parked on an old village site.

Back on the road again, we saw a baby Meller's mongoose, popping his head up out of a hole in a log. A hyena walked by; we enjoyed watching him for some time. A bull elie was picking up tamarind fruit, throwing them into his mouth, and didn't want to move out of our way. Two hippos meandered along together through the bush, oblivious to our presence. As we drove, we saw a lioness strolling, a porcupine in the distance, a tiny elephant shrew next to the road, a genet and a civet running through the bushes, vervet monkeys in the trees, a white-tailed mongoose slinking, a bush hare hopping, and a group of giraffe striding. The only thing missing were the ten lords a-leaping!

An AMAZING evening!!

Back to our magical wonderland, the dinner table was elegantly set, with tablecloth, napkins, china and glassware, out on the grass under the stars, surrounded by softly glowing lanterns.

Vegetable soup
Papadum
Rice
Veggie or chicken curry
Chutneys, onion, tomatoes, peppers, eggs, raisins
Baked bananas
Poached pears

After a wonderful night's sleep, we awoke to a lion call at dawn, and then watched the sun rise over the Manzi riverbed from our room, while listening to the dawn chorus of birdsong. Perfect.

I sat on our second bed to hand Jim our bottle of eye drops; I'd turned back the covers the previous evening so that I could do my leg exercises before bed, and now the sheets were apparently covered in BAT POOP, since this goo was now smeared all over the backs of my thighs! Life in the bush is never dull …

We went for a morning game drive with Phil, and enjoyed crowned lapwings next to crowned cranes, elies in a meadow, and impala with gorgeous early morning light shining against their coats. We took photos of a group of lourie birds in a tree, with their incredibly tall crests.

Before long we came to a lovely rocky glade – I loved seeing the huge boulders! I noticed a plaque on one of the rocks. Phil explained that this was the burial place of Johnny Uys, who had been chief warden of Zambia and Norman Carr's right hand man. Johnny had helped to set up Kafue National Park and was unfortunately killed by a lion in 1973. Phil became a bit emotional, and said that we should visit Kafue Park someday. I sent a silent "thank you," to this man who had contributed so much that we would have these opportunities to see the wonders of Africa.

We spent a long time in the glen, watching impala leaping over the rocks, and a whole troop of baboons grooming, fighting, playing, mating … babies on their mother's tummies or backs, or playing "king of the rock," or leaping through the trees. So interesting. So peaceful.

We watched a hippo wallowing in a muddy stream, with three oxpeckers on his back. Each time he rolled, they would fly up and then come back to land when he settled down. Crowned cranes foraged nearby, and vervet monkeys grazed on the flowers of a wild mango tree. We had to alter our intended tea-break spot due to an elephant herd, but then in our new parking place we also had an elephant "problem." Just after Ellen and I had gone to the loo behind a termite mound, a large bull elie whom Ellen had seen through the brush, walked right up to that very spot. Phil encouraged him to leave by banging on the metal of the jeep and grunting "huh huh huh." He even tried "go away!," but the elie wouldn't leave until HE wanted to. First, he placed himself against a large palm tree and shook it; the unbelievable sound of the leaves shaking and the nuts falling all around us, it sounded like a windy hailstorm. After that display, the elephant felt ready to leave, ears flapping.

Back at camp we had our delicious brunch.

Herb bread
Cabbage-wrapped Indian curry
Beef casserole
Sausage
Lentil cakes
Mixed veggies
Twice-baked potatoes
Green salad
Fresh papaya

I wrote in the visitor's book a song that I'd been working on during our drive:

There's a bright yellow haze in the bushgrass,
    There's a nice yellow sun on the impala;
Cape turtle doves sing,
    We watch crowned cranes and lapwing;
And the lion called at dawn
    While Ron filmed everything.

Oh! What a beautiful morning,
    Oh! What a beautiful day;
I've got a love for this Africa,
    That's so strong I could stop here and stay.

There's a small rocky glad near the meadow,
> Honoring a very special fellow;
We watched louries and impala
> Baboons by the hour;
And the vervets in wild mango trees
> Eating the flower!

Oh! What such wonderful safaris,
> Oh! Thank you for the marvelous memories;
Candlelit dinners, under stars, Lions calling and mating,
> For next year's visit we all will be waiting.

Phil's legendary walks were stupendous,
> Vernon gave us drinks and the dedicated staff fed us;
Isaiah spotted and protected,
> Babette's hospitality is perfected;
And the night drives with leopard
> Were really quite wondrous.

Oh! This camp is a beautiful safari place
> Oh! Each morning begins a beautiful day;
I've got a wonderful feeling,
> We'll be coming back by this way.

Ellen and I "shopped" for presents, sifting through all of the wonderful batiks that Babette had stored in her trunk of goodies. I took photos of each of the staff members; again they loved posing while doing their jobs. I, in turn, got the inside view of the kitchen and gardens and was given a fresh new sausage fruit. It is simply incredible to realize that all the food we'd been provided, dishes more delicious than any we find in fancy restaurants at home, had been prepared by a few guys working with ovens in the ground, and metal plates over pits filled with coals. Jim printed out the photos; they really loved taking their group photo and laughed and pointed at each other in the finished product.

Our jeep arrived, and James drove us to Bilimungwe Camp ("bilimungwe" means chameleon). As we entered the section of the mopane forest swarming with tsetse flies, I suddenly realized that there was no can of "DOOM" in the vehicle! I asked James to stop so that I could dig out a container of Deet from our luggage; even though we knew the flies weren't really affected by Deet, I figured it was better than nothing.

Strangely enough, about fifteen minutes later we came to a fork in the dirt road, and, lo and behold, a can of DOOM sat proudly in the middle of the intersecting pathways. In the middle of the wilderness. James had quietly radioed ahead to another camp and asked them to put a can out front, so that we could pick it up along the way. I was incredibly touched. And I clutched my precious can for the whole drive, using it whenever one of the evil creatures appeared. I only got four or five more bites, and the sausage fruit helped them stay under control as always. Once again, I seemed to be the only one being bitten. Perhaps there is a genetic component to who seems more "delicious" (as with mosquitoes), or perhaps I was simply too attractive by virtue of my constantly fretting and waving my arms.

We stopped on an open plain to watch a puku couple, as James said they were about to mate based on the fact that the male was standing, waiting patiently next to the recumbent female. Interestingly, this male had a dark brown patch on his neck, which came from a secretion from the preorbital gland near his eye, smeared down his neck with his foot, and signifying that he was the dominant male in the herd. After waiting for a few moments, he began trying to encourage her interest in mating by gently tapping – actually, kicking – at her. Finally she stood up, but kept eating bits of grass and giving half-hearted alarm calls, suggesting that she wasn't particularly interested in mating. We watched for many minutes, while he valiantly tried to mount her each time she stopped eating, and his tiny equipment would be ready to go ... but she'd slowly walk away and his front legs would fall back down to the ground. We finally left them, hoping that eventually his hard efforts would pay off.

Our next encounter was with a large group of about ten elephants, moving across a vast marshy lagoon from left to right in front of us. We watched them playing in the mud, and then their tossing sand baths. I marveled at the two babies in the family, as they barely managed control of their tiny trunks. A little further along, we came across another family of about ten elies, also moving across the road.

We turned down a side track, caught sight of a fork-tailed drongo, and arrived at our new camp.

Nigel, the host, showed us to our chalets, which were the most substantial structures we've stayed in yet. They were mud and thatch huts with four solid walls, but with large windows along the sides covered with screens, and a floor made of small stones set in concrete. There was a slate shower with high windows, and a high window over the double tile sinks. Two queen beds and several trunks on which to set our gear furnished the room. Pretty curtains carefully placed over each window were made from local natural materials. We quickly removed most of these so that we could feel as if we were closer to the bush. It was convenient, again, to have the second bed. I asked Teddy, our valet, to put down both mosquito nets at night so that I could read and do my leg exercises late at night without insects bothering me.

The main sitting room was an amazing architectural feat: a soaring vaulted thatched roof, swooping in unusual angles, mounted over an open large wooden deck cantilevered over a small bright green lagoon. It has no walls at all and thus is quite open; but comfort and intimacy are maintained by the smooth dark wooden floors, the two immense tree trunks incorporated as corner posts, the dining table, reclining couches and deck chairs, and complete bar made from a hefty log.

We enjoyed a nice cold Mosi to celebrate our arrival, then drove over to the Luangwa, which is quite close as the crow flies (we could hear hippos guffawing from there each night), for sundowners. It was a PERFECT moment, sitting on the sandy banks of the river, looking down at hippos yawning, and listening to their internecine relationships. Generous portions of Amarula flowed, and Jim bonded with one of the guests, Steve, regarding telephoto lenses.

James taught us about hippo sounds: they start with a high-pitched "whoooo!," which are made when the nostrils are closed, then the following, deeper repetitive grunting "huh huh huh huh huh," results from open nostrils. Jim decided that they sound like a Frenchman laughing.

James and I were chatting, and when he asked which camps we'd visited he mentioned that he knew Patrick, one of our guides from Mwamba camp a few years back, and so I told him of our "surrounded by lions" story. Momentarily distracted by this interaction from watching the river, I MISSED it when the guys sitting on the sandy bank exclaimed that they saw a particularly weird bird. It seemed as if it were two birds hooked together. James and I hoisted up our binoculars and pointed at the dark, strange shape that we could see winging overhead. He proclaimed that it was a PENNANT-WINGED NIGHTJAR! This is one of the birds I've ALWAYS – and I mean ALWAYS – wanted to see. It was gone before I could get my binoculars trained on it.

Jim had gotten a "sustained, detailed" (his words) look.

I'll never live it down - guess we'll have to go back, so I can have another chance. The challenge with this bird is that it only grows its 28-inch long pennants at the end of the dry season, so that it's very unusual to see one before October. We probably won't ever go to southern Africa in October ("suicide month," remember?), so I thought I'd never get to see one. Here was an amazing out-of-season occurrence, and I missed it.

I did, however, see the Mozambique nightjar which was sitting in the road, bringing our total of bird species sightings on this trip to 98 thus far.

On our night drive we saw four elephant shrews, countless bush hares, a civet, several genets, and a white-tailed mongoose. James spotted a gecko in the leaves by the side of the road, hunting insects. HOW could he spot a 6-inch long creature in the leaves in the dead of night? Wow.

Surrounded by gullies and ravines, we parked the jeep on top of an open dirt clearing as there was a known hyena den underneath. Some of the youngsters came out to investigate us, and it was exceptionally thrilling to see an eight-month old hyena baby walk right up to the open side of the jeep, and peer curiously up at us without fear.

The next excitement was a stream of Matabele soldier ants crossing the road, particularly interesting to Mark, a guest and zoologist lecturer at Trinity College in Ireland. He could barely contain himself, begging for permission to leap down off the jeep and examine the ants up close, taking photos from just inches away. It was sweet to hear his technical description – with awe in his voice – of the astonishing abilities of these ants. Apparently they were out hunting for termites and would stop for nothing in their way; when they found the termites they would make a huge warrior-like sound, like Shaka Zulu!

We stopped to watch a family of giraffe, including three tiny babies, quietly striding along in the dark. We learned that their gestation period is fifteen months, and that at birth the babies are simply dropped from a height to the ground! They can actually break a leg in the process. The creatures were magical in the dark, lit up slightly by the spotlight. Lining the background, we could see green impala eyes.

Back to camp for drinks and story telling on the deck by lantern-light, a tremendous dinner, and a night filled with impala alarm calls, a leopard walking right beside our chalet and chuffing loudly, incessant hyena calls, and hippos bellowing wildly. Fantastic!

Marinated vegetable tower
Baked chicken
Butternut squash
Zucchini
Grilled vegetables
Rolls
Peach cobbler

After delicious fire-toasted bread for breakfast, Ron and Ellen went on a walking safari but Jim and I decided to hang out around camp. My tsetse bites were still rather irritating, and Jim was feeling as if he wanted to organize all the photos we'd taken so far, and spend a bit of time relaxing. It turned out to be a wonderful morning! We sat in the deck chairs overlooking the lagoon, and watched the bush life unfold. In the early hours, there were impala drinking quietly from the edges of the hyacinth-covered pond, and an entire troop of baboons in the immense ebony trees on the opposite bank. It was hard not to keep our video camera constantly running, as I kept seeing amazing baboon behaviors: tiny babies clinging to their mother's tummies, nursing, or wobbling on their first steps and trying valiantly to climb up small bushes; older babies playing roughly with each other, pushing and shoving and dragging each other around by their tails, clambering up tree trunks and leaping from branch to branch; adults grooming the babies and each other, or shrieking disciplinary comments to the rambunctious youngsters. One huge male was in charge of the troop, and we could hear his louder and deeper "final say," in some of the altercations. And yet, a few times we watched him tenderly hold and caress one of the babies.

At one point we saw the male mate with a female who was nursing a baby at the time; she held herself in the crouched position with one hand on the ground, and reached around with the other hand to pull the male closer while looking back straight into his face. Very sensual.

We took a break from baboon watching to have warm showers and enjoy our room for a while. Jim was wearing a handsome copper bracelet; the night before, he'd admired it on the wrist of Prince, our spotter, and this morning Prince had given it to Jim as a gift. So sweet.

We came back to the main sitting room and Juwa offered us one of his drink concoctions: pineapple juice layered with a small amount of bitters and then stirred, quite refreshing.

It was fascinating to notice the complete change of attendees at the lagoon. The baboons had all left, and the impala had started a rutting behavior we've never seen before. The four males were all chasing after the one female who was still in estrus (the twenty other females have already gotten pregnant), each male desperate to gain her attention. They were chasing each other at high speed, running full out, with their brilliant white tails held high, and barking like dogs. Simply incredible!

A single family of bushbuck with a baby tentatively wandered down to the lagoon to drink during the impala fracas. And then, a few moments later, the impala were gone too. Several birds flitted down to the lagoon to drink, Meve's starlings, red-billed buffalo weavers, and a lourie. Nigel explained that this little mixed flock of birds came to drink every day around the same time.

The warthogs arrived next, a family of five, drinking and wallowing in the mud and crawling around on their front knees, scrabbling for food in the short grasses. They were very skittish, anxiously leaping up at any small sound, to stand at attention with tails erect, peering worriedly about.

Near the path from the sitting room to our chalets, a group of birds and tree squirrels in a small shrub screeched and chattered at the ground underneath. Nigel investigated and discovered a puff adder quietly hiding under the leaves. Yikes! The first snake we'd seen in Africa, the puff adder is highly poisonous. We asked Nigel to please remove the snake from the camp, feeling that there was plenty of room out there in the rest of the bush for him. Nigel told us that he did carefully remove the snake later, using long branches to pick him up.

The walking safari gang returned in time for a delicious brunch on the deck and then a relaxing siesta before our afternoon activity with Manda, who had recently been described by "Getaways" magazine as one of the best safari guides in Africa. I collected pretty nuts near our hut; they are called "air potatoes" as they look like potatoes on the inside, but are very light-weight (they grow on a vine native to Africa). Before leaving on our game drive we saw that the species hanging out at the lagoon had shifted yet again: now it was the reign of the tree squirrels and a new mix of birds: blue waxbills, fork-tailed drongos, and dark-headed bulbuls.

Sausage and prunes wrapped in bacon (aka, "angels on horseback")
Baked beans
Spicy veggie pizza
Baked marinated mushrooms
Dijon bream
Pasta with snow peas
Cabbage salad with tomatoes
Cheese bread
Custard cake with raisins and chocolate icing

The afternoon had become windy and a bit cloudy, but this weather system luckily only stayed for an hour or two. Manda taught us about the white-browed sparrow weavers, who are "cooperative breeders," whereby only one alpha male and female actually mate and lay eggs. The rest of the flock are simply involved in helping to build the nest and raise the young. He showed us a nest which had blown down in this wind, and pointed out that they have two entrances: one is quite obvious while the other is hidden. In the case of a predator entering via the noticeable entrance, the birds can sneak out the back door to safety.

We saw an impala with a sprained knee, and understood that a mopane forest was a clever place for the injured animal to hide in, as he could hear a leopard sneaking up on him due to the noisy fallen leaves underfoot.

A few elephants browsed nearby in a thicket, and Manda taught us how to distinguish male from female based only on their heads (if one couldn't see well enough between their legs to "check their documents," as James had suggested). Females have an angular forehead while males' are more rounded, and the tusk attachment point is broader in males than in females. Interestingly, tusks grow forever, and thus the bigger the tusks, the older the individual. However, elies' teeth do fall out and are replaced: they have six full sets of teeth to use over the course of their 50 to 60 year lives.

On the bank of the Luangwa River were nesting holes of hundreds of carmine bee-eaters. The birds were flitting out of their river bank burrows to hunt for insects at dusk, but they all first landed on the sandy bank of the river to sit for a while, perhaps to warm up before scrounging food in the cool overcast. These birds are a gorgeous brilliant scarlet and blue, with long racket tails pointing out behind, and they looked quite pretty sitting lined up on the sand in a large flock. As migratory birds, the carmine bee-eater returns every year to the same location to nest.

We enjoyed our sundowners on a grassy open field, surrounded by impala and puku. As the sun set, hyenas arrived. Ellen was able to spot them far away; they looked like stumps to me. An open field is an excellent place from which to listen for alarm calls, which the hyena would then follow to find a leopard or lion kill. They actually take kills from leopards, but have to wait their turn at lion kills. We listened to a young Pel's fishing owl calling for food - it sounded like a person disappearing down a bottomless pit, quite plaintive.

Our first nocturnal sighting was a bushbaby, its tiny dazzling eyes shining down at us from up in a tree. It's unbelievable how the guide and spotter could SEE these little guys in the dense foliage. Later in the evening we saw two more, although I have to admit I couldn't distinguish the outline of their bodies even with my binoculars.

I kept track of the genet sightings, and we saw seven this night. One was running extraordinarily fast. Manda also found another puff adder on the side of the road, much smaller than the one we'd seen in camp. The smaller ones are actually more dangerous, however, as they will strike at anything, being more anxious than the more experienced snakes, and dump all their poison at once rather than metering it out based on size of its victim. Puff adders are responsible for most snake bites in Zambia, and perhaps even all of Africa. The limb which is bitten swells up astronomically within minutes, and sometimes there can be so much necrosis that the limb needs to be removed. Manda said that there are about 40 species of snake in Zambia, only "10 percent" of which are poisonous. However, when he started listing the poisonous ones (black mamba, spitting cobra, vine snake), he listed at least eight; all of us on the vehicle laughing amongst each other about the incongruity of these numbers with the claim of "only 10 percent." Most bites occur at the beginning of the wet season, as there are more insects about and therefore more snakes come out.

We saw a bush hare, a civet, and a water dikkop (the "dead battery bird" based on his long, descending and slowing call). We also noticed the bright eyes of wolf spiders sparkling at us from the brush.

And then, when we were almost back to camp, we saw it.

Our second LEOPARD!

He was peacefully lying in the road, and we were able to get a detailed view of his rosettes and luxurious coat. We could even admire his face as he disdainfully looked our way, eventually standing up and sauntering off. Seconds after we began to drive again, we came upon two giraffe, also in the road, and they gawkily loped away as we approached. Only seconds further along, at the actual entrance to the camp, were four elephants, also IN the road. What a night! We had to drive around, through the brush, to make it into camp.

Creamed mushrooms on toast
Rolls
Filet of beef
Gem squash filled with butternut, corn, and potato
Carrots glazed with herbs
Cauliflower in cheese sauce with red peppers
Baked potato
Poached pears and apricots in wine with orange peel

After this gastronomic feast for dinner, Jim and I retired to our room earlier than the rest of the gang, who were chatting and laughing over after-dinner drinks. We heard the leopard chuffing, a baboon bark, and a few puku alarm calls. Suddenly we heard a very loud ruckus, and I called over to the main room to ask what the sound was. "What do you think?" Manda asked.

"Mating leopards?" I excitedly suggested.

"No, fighting hippos," was the answer. Perhaps I'm STILL not ready to be a safari guide.

We fell asleep to the sound of elephants in the bushes next to our window, and I was awoken in the night by a bat swooping over my face, inside the mosquito netting. He'd literally touched my face with his wings! I was scared, but Jim let him out for me. My hero.

When morning arrived, the first thing I said to Jim was, "What time is it?" and after his answer, "5:15am," I happily proclaimed "Hares, hares, rabbits, rabbits," only to have him comment that I'd already said something when I asked him what time it was. Nigel had told us the previous evening how his grandmother had always insisted that on the first day of each month, one needed to say "Hares, hares, rabbits, rabbits," as the first utterance, in order to have good luck for the ensuing month. Guess I blew it.

So then I followed up with, "HAPPY BIRTHDAY!" He was turning 51 and still as gorgeous as the day we met in 1971!

Our morning game drive and walk with Manda was as great as usual. We saw an African hawk eagle on a tree with his kill, and watched the eagle's head turn 180 degrees like an owl. Our scout, Gabriel, got out of the jeep to remove some elie damage from the road (broken branches) and soon thereafter we came across said elies … a mother and a baby … and received a loud trumpet. A flock of Lillian's lovebirds flew overhead, looking like a shimmering green carpet. We enjoyed the sight of a white-fronted plover taking a bath in the river, and then parked the vehicle in the shade and took a walk in the bush. We examined mongoose and elephant shrew tracks and myriads of termite tunnels. Known as the "dust bins" of the bush, they engulf any vegetable matter and digest it with the help of symbiotic fungus. Termites need these mud tunnels to protect them from the sunlight, as they have no pigment and a high surface area to volume ratio, and can therefore easily dry up if exposed. Their mud plasters leading up the trunks of trees doesn't harm the tree, as the termites only feed on dead branches and the outside layer of the bark.

We learned about the village uses of various trees:

- Women smoke the seeds from the trichoderma tree to attract men; they'll even mention the name of that special someone as they smoke, hoping to influence subsequent events.
- The fruit of the monkey orange tree can be used to block the neurotoxin from the black mamba, but beware of eating the fruit as it contains strychnine before it ripens.
- The roots of the silver terminalia tree are mixed with blister beetle entrails to produce a strong diuretic for flushing the system. Wasps lay their eggs near the branches and inject a growth hormone so that a "gall" grows and the larvae can thrive.
- The green peeling bark of the thorny zebra wood tree is used as poison for darts.
- Women use a tea made from the bark of the scrambled egg tree to terminate unwanted pregnancies. As this is often done in secret by young girls who don't find out the proper dosage, it unfortunately often only leads to problems for the mother and the child rather than ending the pregnancy.

A group of ants with grass seed husks piled up outside their holes prompted Mark to get down on his knees and pick one up to show us, calling to the ant, "Come here, I'm not scary!" We also discussed hippo territories, finding out that a particular male may only control a small area in a river, but has command of a much larger section of the bush next to the river, where the foraging occurs. They even tolerate other males in their landed territory, whereas they would never allow such a trespass in the water. The hippos eat sausage fruits in October when their more preferred grass is scant, and so their "highways" often lead from one sausage tree to the next. They can eat fifteen sausages a night, quite a feat considering how big these fruit can grow. From September to December tremendous hippo battles ensue as food supplies become less plentiful.

As we examined an aardvark's burrow, Manda explained that they come out for only a few hours at night to eat ants and termites, but in that short amount of foraging time they manage to eat up to 70 kg in weight. That's a lot of bugs. We examined a buffalo head skull which had been invaded by "horn moths." These critters push their feces out from the horn in a long tube, down which their larvae eventually crawl when they are ready to fly. The ash from burned feces tubes can be applied to rashes. I'm not sure I'd be willing to collect the tubes to burn, however, as they looked really alien and gross, sticking out of the horn like medusa's locks.

When we looked at an old hippo skull, we learned that one ought not to pick up old skulls from the ground, as anthrax kills many animals in Africa and the spores could still be present.

Sautéed potatoes and onions
Butternut/pumpkin tart
Ham with deviled eggs
Avocado/tomato salad
Pear/tomato/mushroom salad
Balsamic onion bread
Fresh papaya

After brunch back in camp, we took photos of the staff, and as always had a fantastically fun time. The staff worked hard to get dressed up in their best camp clothes, and I was invited into their compound to take photos of Sam ironing our clothes with a flatiron into which he would place hot coals. "Come into my office!" he invited, then joked with an engaging smile, "Oh, my computer is down today," while Joseph and Alfred posed cutting up vegetables and also putting fresh bread loaves into the stone oven. Teddy wanted to be in our hut, making our bed and smiling widely. Juwa posed with a book in his hand and a pair of binoculars (he has aspirations to become a guide when he has finished providing money for his brother to complete HIS schooling), then asked for another photo catching him serving a drink. Gabriel posed with his rifle, very serious in his scout uniform, and Prince proudly stood next to the jeep. They watched with interest as Jim printed the photos, and graciously accepted the photo print gifts with great excitement.

It is a tremendous honor to me, that we are able to provide these deserving, ambitious, and hard-working guys with needed jobs. They obviously appreciate our being there, too. Furthermore, I feel that one of the most important aspects of our being in these safari camps in Africa is the better understanding that we each obtain about each other's different cultures and homelands. And this understanding will lead to a better world for us all.

Back at our tent I took a quick shower. Just as I covered my hair with shampoo, a family of elephants arrived at the lagoon. I jumped up and down, trying to see out the high window, as the lagoon was right behind our chalet. Jim ran outside with a towel wrapped around his middle, peering around the side of the building and snapping photos. I rinsed off the soap as fast as I could, and came out in a similar towel, with video camera in hand. It was a peaceful sight, the family of four elephants sucking up the water with their trunks, shooting it into their mouths, and then wandering off into the bush when they'd had their fill.

Our afternoon game drive was filled with excitement. We found "luck beans," (tiny seeds that are bright red with a little dot of black on the end), which are sought after by village women as the luck beans "make them hot for men." We learned that the lead wood tree leaves could be pounded into a powder and used for wound healing and as an antiseptic, and that the natal mahogany leaves could also be ground and used to cure the rash caused by the false coffee-bean tree.

We saw two crowned cranes, with a juvenile – it was very exiting to see a young one as they are a highly endangered bird species.

Our "usual" hippo spot on the Luangwa served as our sundowner spot this evening, and we enjoyed extensive and breathtaking views of the babies. A group of four was playing with each other in a circle, each one practicing yawning, as they will eventually use this behavior as they seek dominance within their group.

On our night drive, we watched a hyena wandering and a puku pronking; a genet in a tree and a mother and baby elephant by the road.

And then we saw them.

Our third AND fourth LEOPARDS!

They were walking across a muddy lagoon, and we watched them with the spotlight for a while, as they elegantly strode across the dirt. Prince turned off the light for a moment to let them have a rest, but then we heard a ruckus and turned it back on in time to catch our first leopard chase as the male pounded after a white-tailed mongoose. The mongoose was unbelievably fast, especially over the bumpy muddy meadow, and the leopard gave up pretty quickly. The mongoose was quite a sight, with his white tail pointed straight up in the air as he jinked and jagged away from the pursuing cat. After this thrill we watched the leopard take a long drink from the lagoon, and he then sat down, licking his paws, fur, and lips.

Jim joked with me on the way back, that he wished the leopard had gotten the mongoose, so that there would be one less mongoose in the world for me to mistake for a lion. Very funny, Jim.

Back from the drive, after Juwa served us drinks from the bar, Nigel invited us to "Joe's Bistro," a very sweet outdoor dining room that the cook, Joseph, had built for the camp in front of his cooking hut. There were low grass walls around a sandy central courtyard where the dining table was set with white linen, china, stemware and candles. A fire burned in a small side enclosure surrounded by deck chairs while reed decorations looped around the trees and lanterns shone warmly. I cried again! After cocktails by the fire, we devoured our delectable dinner served out under the stars.

Butternut squash soup
Nshima (typical Zambian dish of ground maize or mealie meal)
Lamb stew
Vegetarian burritos
Eggplant with tomatoes
Cabbage/raisin/nut salad
Fresh bread

After dinner, the staff caught us by surprise. They carefully walked out from the cooking hut, blocking the wind and carrying a stunningly beautiful birthday cake (for which they'd had to send safari jeeps around to other camps to get all the ingredients!), decorated with candles. Sam had drawn multi-colored icing designs all over the cake, honoring Jim's birthday. With radiant smiles and glowing faces, the staff sung, "Happy birthday," in a Zambian tribal language. As Jim blew out the candles, we were all crying.

The cake was delicious, filled with pieces of fruit as Nigel had secretly discovered that Jim didn't like chocolate cake, so they cleverly improvised. And I sang a song I'd composed during our night drive:

Happy birthday to Jim,
    A safari in Africa for him;
Elies, leopards, and hyenas,
    Amarula to the brim.

Nigel makes it all great,
  Juwa's service is first-rate,
Teddy cleans our tents perfectly,
  Gabriel protects our fate.

The best cooks anywhere,
  Are Joseph and Alfred, that's clear;
Prince spots like a leopard,
  It's a joy to be here.

We saw a chase by kaingo*
  Manda led us to the hippo;
Lagoon with warthogs and baboons,
  From Bilimungwe we never want to go.
                              *leopard in Bemba

Suddenly, we heard the loudest sounds we've ever heard in Africa. In a tree only about 50 yards away, behind the cooking hut, were about a hundred baboons, all of which were shrieking and screaming and bellowing. WOW!

Just then, Manda exclaimed, "Leopard!" while we all stared in the direction of the clamor. "Let's go see!" he cried, pointing towards the jeep. The four of us ran after him along with Gabriel and Jason (the driver sent from Mfuwe Lodge, who was to drive us out of the bush the next day). We had no sweaters, no boots - just flip flops, no cameras or binoculars, and Ron had only a short bit of video tape left in his video camera. It was a very intense experience.

We drove out the track around the other side of the lagoon, and arrived under a towering ebony tree filled with the screaming and yelping baboons. After a few moments of scanning the branches with the spotlight, we saw the leopard in the tree. Earlier in the day I'd claimed that I had begun to believe that leopards never really go up in trees, as we've never seen one do so. Well, here he was, our fifth leopard!

And what a gorgeous creature he was, a gigantic male, with a gloriously lush coat and massive head. Next to him in the tree was his prey, a dead baboon hanging upside down, with blood streaming down her face. It looked very human, and I have to admit that the leopard looked quite sinister to me after I saw the dead baboon's bloody expression.

And then we saw something even more incredible. There was a baby baboon on the ground under the tree. It was only about three weeks old and must have been dropped by its mother during the kill. It had fallen at least

50 feet but was still alive, clumsily clambering over the leaves and twigs, plaintively calling and whimpering.

All of us were touched by the pathetic movements of the poor baby. We knew that another female would happily take over raising an orphan, should it survive its injuries, but there didn't seem to be any way for the baboons to come pick up the baby with the leopard around. The baboons approached the leopard several times en force, shouting directly at his face, but he didn't make a move. Normally the cat would have left his prey in the tree for a while before consuming it, and we guessed that perhaps the presence of our vehicle was preventing him from leaving the tree. We therefore drove back out from under the branches, and watched from a distance. This helped apparently, as the leopard soon slunk down out of the tree, with surprising grace even with a bloodied leg, as he delicately picked his way through the branches, down the trunk, and off into the bushes.

After he left, the baboons screamed even louder, if that was possible, as one by one they would start to come down the trunk of the tree towards the ground, only to get frightened by the possibility of leopard attack and rush back up to the high branches. Even the gigantic male tried his luck, grunting amazing low guttural sounds as he tried to come down, and shrieking during his retreat. We moved the jeep even further away, circling through a thicket, thinking that perhaps the jeep was bothering them. But apparently it was the leopard, not the jeep, as when we got back from our circuit the baby was still squirming and crying, and the baboons were still hanging on the upper branches, all looking down at the infant and yelling.

It was just too much for Gabriel, and he stepped out of the vehicle, rifle in hand, walked over to the baby, reached down to pick it up, and came back to the safe haven of the jeep. This made us all cry out with pity as the baby struggled and whimpered in his hands. The baboons were still shrieking, and when Manda told Gabriel to go over to the tree to place the baby on the termite mound at the bottom of the tree, he was incredulous. "I'm not going over near those baboons!" he exclaimed. So, Manda drove the jeep towards the tree as Gabriel walked next to it, and when we reached the termite mound he quickly deposited the baby and jumped back in the vehicle. We immediately backed up, and a female baboon swooped down out of the branches and tenderly scooped up the tiny baby.

Ahhh! We all heaved a sigh of relief. What heroes we were.

Back in camp, we had a long night, listening to the baboons' never-ending shrieks and replies, while a lion called quite close as well. By morning the leopard had killed a second baboon, explaining the night-long vigil.

In the morning we left after breakfast, and as usual, it was hard to say good bye to the wonderful camp. Manda took off for a game drive with Mark and Charlie after we all exchanged addresses. We packed up and climbed into the jeep that Jason had brought for our transfer.

As we drove through the mopane forest which was not quite so dreadful in the cool morning, before the tsetses woke up, I noticed some large splashes of water from the river off to the right. I realized that it was a giraffe walking upstream in the water. I asked Jason to stop so that we could see it more clearly. We'd never seen a giraffe walking in a river before, and it was an interesting sight to see the large amount of water this tall creature displaced with each footstep. Jason carefully looked around through the forest and low grasses, and decided we could all get out and walk over to the river bank to have a closer look. As we watched, we could see Manda's vehicle also stopped far away on the other side of the river. We hoped they could see the giraffe too. Suddenly their vehicle started moving, and drove across the river towards our side, crossing on the submerged sandbags. We leisurely walked back to our jeep and began moving along the road. Only a few moments later, we stopped for yet another amazing sight: lions were walking across the road, from right to left, all around us. There were about six lionesses and one lion, and we quietly took photos and video as they literally strolled within a few feet of our jeep.

Manda's vehicle then arrived from ahead, and as they pulled alongside we queried Mark and Charlie whether they'd seen the cool giraffe in the river. "Yes," Charlie explained, smiling enigmatically. "We could see you on the riverbank, and the giraffe … and also the lions, only a few yards away."

Wow.

We hadn't thought of that, when we encountered the lions only seconds after getting back in our vehicle. Manda had hurriedly driven over to inform us, in case we weren't already back in our jeep, of the presence of the little kitties surrounding us while we out on the river bank.

We began our drive again, and had yet another unforgettable experience: a group of about seven hyenas collected around an abandoned lion kill, quietly munching. One of them turned towards us with a huge bone sticking out crosswise from his mouth, at one end of which we could see a tiny giraffe hoof. Guess we knew what the lions had for breakfast.

We had told Jason that we wanted to collect some sand from the Luangwa River for our collection, and he found a perfect spot with dazzling white sand. While he was out of the vehicle, a nearby giraffe was startled and awkwardly but quickly ran along the grassy river bank. We later saw another group of giraffe, three adults with three babies, and Jason told us that this family had four infants when he'd seen them earlier in the week. The babies were only about four months old, and were very sweet to watch as they approached one another and then walked side by side. The adults loped as we approached, which seems ungainly but somehow also graceful.

Our next amazing encounter was a family of about twelve elephants, with two really small babies. One of the larger females was right next to the road, and trumpeted quite loudly as we passed. She then gave us a great demonstration of how to obtain grass: twining her trunk around a bunch, pulling up while using her foot to assist breaking the bunch free, then thwacking her forehead repeatedly to remove the dirt from the roots before smashing it into her mouth.

We saw a majestic recumbent male lion in the shade along a dry river bed, two intensely vivid black, red, and yellow saddle-billed storks dexterously picking through some bright green grasses, two male buffaloes eyeing us closely, and four female kudu, including a baby, posing in front of a lovely baobab tree. And all this was on our simple "transfer" drive.

We stopped at the Mfuwe Lodge for coffee and a bit of shopping.

Crossing over the Luangwa River bridge, I had a good cry, to say goodbye to the South Luangwa Park for this year.

We then had a pleasant time visiting the town of Mfuwe, hosted by Jason. He walked us through the vegetable market, and bought us corn fritters that we admired, as a gift. He helped me to purchase batteries and ibuprofen in two shops. We hadn't exchanged any money so didn't have any kwacha, Zambia's local currency. Jason kindly paid for me and I repaid him later in US dollars. People were dancing in the streets to reggae music as it was almost election time and as such, a lot of campaigning was going on. A group of children crowded around when I got out our camera, and loved posing and then seeing themselves in the digital display on the back. One little boy in particular was most endearing, and followed us around, with big eyes and a wide smile.

Jason asked if we'd like to visit his home, which was a very special experience. We loved driving on the dirt roads, passing villagers on bicycles carrying huge sacks of ground mealie meal home from the mill. Their clothing was brightly colored, and the women and children carried improbably large loads balanced on their heads. We met Jason's brothers and nieces at his house (his parents were unfortunately not there), and I realized that not only was Jason proud of showing his nice home to us, but he was proud to be showing us, his clients, to his family. He described the three small buildings that they have, in addition to the square mud brick house, including a small room for the fire (to block the wind while cooking), an outhouse with a long-drop loo, and a room for using water in buckets (collected at the local bore hole and brought back in plastic jugs) to bathe. And Jason was very proud to show us the papaya and orange trees he'd planted, as he would sell the fruit to the safari camps and raise extra funds for schooling for his brothers, who told us they aspired to become guides just like Jason.

We had our delicious "box lunch" in the courtyard at Tribal Textiles, where we bought yet more gifts to bring back home. We love their batiks, and are pleased that the local company has become quite successful with their sales to the safari business.

At the airport we had Mosis at the Moondogs Café, paid our $5 exit fee and walked through the beeping metal detector with all our luggage, including Jim's knife, as the attendant simply waved us through. We took off from the airport in a comfortable Chieftain airplane where Jim sat in the right seat up front and was quite happy.

I started my serious crying when we arrived at the Lower Zambezi Park and I looked down to see Jeki International Airport, remembering that moonless night on one of our previous Africa trips when I had been thinking of Kalpana all the while Jim held my hand. We flew over Sausage Tree Camp which brought back fantastic memories of our two visits there. I could see the lagoon waterways we'd canoed through last year, surrounded by hippos. I could see forever - Chiawa Camp, Chongwe Camp, and CLZ. This land was becoming so familiar to me that I even particularly noticed the spot where we'd seen the black heron fishing last year. Jim later told me he too had looked down and noted all the same special places.

We arrived at Chiawa with hugs for Joe and Barbs and Craig, while I cried incessantly. We gave Barbs all the gifts we'd brought from her sister and the radios for Grant, and settled into our tents. These chalets were the traditional large safari tents mounted on wooden platforms, looking out on the Zambezi River, with open-air loos out back.

The chalet was cozy with a nice wooden headboard behind the bed, an open set of shelves, a small desk, and luggage racks for ease of storing gear. Out front, the deck had several reclining chairs and a low table. Chiawa was much bigger than our previous camp, with nine chalets instead of four, catering to more guests. The main room had a large bar/sitting area and a separate room for dining. All of these rooms had only a few sides made of thatch, opening completely towards the sloping grass leading down to the riverbank.

After getting our luggage settled, we took off for an abbreviated afternoon game drive with Joe as guide and Wallace as spotter. As we slowly drove away from camp, about five minutes into the drive, Wallace pointed excitedly at a bent-over tree trunk right next to the road.

And we saw it.

Our sixth LEOPARD!

He was a young male, about two years old, and was hiding under the tree in a crouched position, staring out at a group of impala in a nearby field. We had a great time watching him … he eventually came out of hiding, stalked his prey, freezing at times and crawling at others, until eventually giving up on the impala and wandering off into the thick bush. It was wonderful!

We'd missed actual sundown being as involved in watching the leopard as we were, so Ellen quipped that she guessed we were thus going to have "sungoners" tonight. Joe drove us up to a spot on top of a ridge, and we could see the dark outline of the escarpment against the purple sky, and the golden remnant of the sunset. Moon was overhead, becoming quite full, although we could still discern Jupiter and the Southern Cross pointer stars. Wallace set up a wooden table, complete with white table cloth and candle, and we had our Amarula and veggie samosas. The scent around us of the woolycaper bush was intoxicating. Joe picked a few flowers for Ellen and me. We saw a nightjar and a water dikkop, while a young male hyena seemed to actually follow us around. We also noticed some bright spider eyes. We saw a giant eagle owl perching in a tree, and were able to appreciate his pink eyelids before he flew off. Joe explained that owls hunt mostly by sound (their ear muffs funnel sound to their ear holes), as their eyes actually only have a 2% greater aperture than ours. During the day they hide in the winter thorn trees, where the forked-tailed drongos bombard and pester them to leave.

About 8pm, we headed back to camp for drinks around the fire on the bank of the river, and then the Chiawa choir sang for us.

I cried. And Ellen cried. I was so excited to have been able to share this thrill with her. I love the sound of the staff singing and clapping in the African night.

Veggie croquettes
Rolls
Asparagus soup
Beef filet or crusted bream
Potato slices
Broccoli and cheese
Flaming Crepes Suzette

At dinner our wine glasses were refilled so frequently that we all got quite tipsy, and subsequently a little silly, all good fun. During the night we heard explosive hippo sounds – my favorite – as I snuggled with our splendid hot water bottles.

We were woken next morning by the deep sound of hippo calls right before our "good morning" greeting, and tray of coffee and tea delivered to our deck. We actually watched the sunrise over the Zambezi from our bed, while drinking our morning beverages. Paradise!

It was rather chilly, so I wore my hat and gloves for the first time on the trip, to breakfast on the riverbank. As the sun rose higher in the African sky, I warmed up quite nicely.

Our game drive with Joe started with a female bushbuck, whose fur was fluffed up so that the sun could penetrate and help her keep warm. It was a brilliantly clear day, with breathtaking views of the escarpment. We decided to go on a lion hunt, traveling quickly downriver to a spot where a large lion pride had last been seen. Joe reminded us that lion tracks have three lobes at the back of the pad, while non-cats have only two. Jim put on his Rommel-style goggles for traveling fast in the open air, and we were off.

We saw amazing leaping impala, so magnificent that they took my breath away. Groups of trumpeter hornbills passed overhead, crying like babies and stopping to eat figs from the strangler fig trees. We saw gorgeous baobab trees alongside star chestnut, aka "false baobabs." We watched hippo in the dambo, peeking their heads up between the green leaves of the hyacinth, with cattle egrets on their heads and jacana stepping nimbly about. The hyacinth is quite pretty, but unfortunately is an invasive from South America and can wreck havoc by choking waterways. No local animals will eat it.

We watched a huge gang fight between two rival troops of Chacma baboons, which are much larger than the yellow baboons in the South Luangwa, filled with contentious yelling, barking, chasing and shrieking babies. Very exciting! One of the baboons near us laid flat on the ground, and Joe explained that he was doing this to "hide" from us - didn't work, though.

We enjoyed an enormous herd of buffalo (and made silly jokes asking each other: "Heard of buffalo?" "No, have you?"), many of them moving amongst the thickets. There were lots of youngsters, and we watched the crowd until they began "moo-ing" away.

I enjoyed the thinner undergrowth in this area, as there is more hard-packed earth and so we could track the animals and spot game more easily. We deeply inhaled the delicious scent of sage as we drove through the bush following lion spoor. Joe showed us many tracks, including those of the local male lion, Douglas, but they were going in too many directions to decide which direction was the most likely to find them.

Two bull buffaloes stared us down, and Joe commented that they "look at you like you owe them money"! These big guys are scared of nothing; they even killed Dan last year, the other male lion in the area.

We visited "hippo city," a small stream where a large group of hippo were resting in the water together, placing their heads on each other's rumps for comfort. We watched from a nearby grassy knoll, where we stopped for a coffee break with snacks. It was so peaceful on that grassy plain, with blacksmith plovers in the dambo nearby. A family of warthog ran from us, and in its hurry, one actually ran into a bush and stumbled!

Back in the jeep, we watched a tremendous herd of buffalo run past us on the road, their footfalls sounding like rain as they stomped and crushed the undergrowth. Another dambo had about forty egrets in the hyacinth, with several hippo heads peeking out and a black crake clambering about.

Joe showed us a "young" baobab (only about one hundred years old), which had a pretty, entwined set of branches. The baobabs flower in October, but they only open at night; the fruit can be eaten or used to make cream of tartar.

Unsuccessful at finding lions, we returned back to camp for a delicious lunch and a cold Mosi. Then it was time for siesta, with fantastic outdoor showers and a wonderful relaxing time in our tent.

Potato salad
Cajun rice chicken
Salad with croutons and cheese
Four-seed bread
Cheese plate with crackers

At tea time we had bruschetta and berry tarts, and then were off for a boat ride on the Zambezi. We cruised amongst the islands in the river, enjoying the views of the escarpment, the gorgeous green grasses on the islands, and the brilliant blue of the water, as the Zambezi is quite clear and not muddy.

We saw a tiny crocodile about one year old, which looked like a piece of wood. We joked that normally, pieces of wood look like crocs, rather than the other way around. White-crowned plovers frantically chased baboons away from their eggs on the ground. We watched a male kingfisher (I could tell he was a male, by his "waistcoat" of black, rather than the "bra" that a female sports) splash-bathing in the river and preening on a branch, and we saw a fish eagle nest with both parents in attendance (normally two eggs are laid, but the first chick born pushes the second one out of the nest, known as the "Cain and Abel effect"). A yellow-billed stork used its wing to make shade, to fool fish into thinking they were safely under an overhanging river bank. And we saw a malachite kingfisher diving for food. I can't imagine what kind of fish such a tiny bird could skewer!

We came to a heron rookery, where gray herons were nesting, and a large number of black-crowned night herons roosting. We flushed them and they flew en masse to a nearby sandy spit, where they looked like penguins as they stood, all lined up, on the sand. We saw African darters, which stab their prey with their pointed beaks, and realized that we'd now seen 127 species of birds on the trip so far.

We enjoyed sundowners on the boat, with Ellen and Ron and another couple, Luciano and Judith, enjoying snacks and taking photos of the sun setting over Chirapira Mountain. We drifted a bit further down the river, and put in at Sunset Point where we joined the jeeps for a night drive. We saw a family of elies, with a very tiny baby, and fireflies flashing in the dark. The brilliant moon showed the "rabbit," or Kalulu, on its face, as we enjoyed the scent of the potato bush. Jim referred to our jeep as "the stealth vehicle," as it had a broken back seat (hence only four of us could be on the vehicle) and was quite noisy. We searched for nocturnal animals but only saw one white-tailed mongoose, and so called it an early night as we headed back to camp. Again, drinks were served while the Chiawa choir sang, followed by dinner under the stars.

Crab salad
Rolls
Zucchini soup
Beef, pork, or fishcakes
Mashed and baked potato
Julienne vegetables
Warm chocolate cake/pudding

During the night, we heard hyena and hippos calling, and then we heard two buffalo eating right outside our tent, scrabbling in the leaves. At one point, one of them actually bumped the tent which gave me a huge fright.

The next morning we watched sunrise again, and had another delicious breakfast around the fire out on the riverbank. As Ron and Ellen were going fishing, Jim and I had the jeep to ourselves, with Daniel as our guide. We proclaimed that we wanted to see birds and elephants, and although the wind suddenly picked up and the sky became overcast, sending most birds into hiding, we still had a fabulous time. We saw crested francolin and learned that they do put up a crest when calling (otherwise we couldn't see it), and Daniel explained why we always see baboon and impala together: baboons like to eat baby impala while the impala benefit from the baboon's better eyesight and alarm calls for predators.

And then, we saw elephants. As we'd requested.

We saw a mother and a two-year old under a tree and as we watched, the baby walked out into a nearby field, quite far from mommy. Unusual. When a vehicle drove by on the other side of the field, he hurried back to mommy's side.

Shortly thereafter, we came across another mom with a year old baby, its tiny little tusk buds starting to show. He was getting proficient at the use of his trunk, and we watched him place one seed pod in the curve of his trunk while he ingested a second one; then he grabbed the first one with the tip of his trunk. He rubbed his bum on a tree, while his mom rubbed on another tree, and then shook the tree to get more seed pods. The baby then reached way up into the tree with his trunk.

Passing by several lone bull elies, we next came to a mom with a juvenile, a two-year old, and a TINY fuzzy baby. The baby was snuggled in amongst the four of them in the bushes. At one point we could see his little trunk sticking up alongside the mom's shoulders, but nothing else … it was so cute, wiggling around.

Driving around to the other side of the bushes we tried to catch this group in the event it wandered through, but instead we saw a very sweet young bull who was very wary of us. He kept his ears out and faced towards us the whole time, as he eventually crossed the road right in front of the truck.

We continued to stalk elies until we found a group of four adult females with a pair of two-year old babies. And the encounter was the most amazing of my life! The babies tussled with each other for thirty minutes while we gazed in awe, filming and snapping photos, or just sitting and taking it all in. They wrapped their trunks around each other, they tangled trunks, and even ended up with their lips entwined at some point. They pushed and chased each other around, spreading their little back legs to push harder, just like grown elies when they push against a tree to knock down the pods. It was marvelous. One time they ran at each other from a distance, butting heads. It always ended in a stalemate, with trunks entwined. The moms ate peacefully while the kids played, oblivious to our presence. Eventually they moved further into the brush and we moved on, and that's when my tears started. I just couldn't stop; my cheeks were sopping wet. I felt so privileged. I just couldn't believe that we could see this in the wild; that we were so lucky that we could be here.

We next came across another young bull, and although he seemed quite relaxed to have us next to him, it wasn't until we moved on that he gave a loud trumpet. I explained to Daniel that I'd always wanted to get an elie trumpet on video, as I loved the sound.

So, when we saw another group of elies down by the river, four or five females with a very tiny baby right next to the road, Daniel drove the jeep up next to the mom/baby pair, and I was ready to film the enraged trumpet, but the mother mildly looked at us and continued browsing. We watched for a while, and she was completely comfortable. We all had to laugh!

It was time to head back to camp, and en route we spotted several large hippos on an island, looking like large pinkish-gray slugs. We came across a young bull elephant hanging out with an older bull, browsing together. A stately elderly bull sauntered next to us without a care.

And finally we intercepted another mom with a tiny baby, who was all covered in mud and who watched us intently, even seeming to try to pick up our scent with his minute trunk.

We'll never forget these 10 intimate elephant encounters. We thanked Daniel profusely as we headed back to camp.

Chiawa had a surprise in store for us. We were led down to the boat launch and invited aboard a pontoon boat for a champagne brunch! A table for six was set and we cruised along the Zambezi while eating our delicious meal on board, an incredibly decadent experience. The setting and china were elegant, even the coffee and tea were served from silver pots, all out on the water. We passed a huge elephant on one of the grassy islands (encounter number 11 of the day), and watched saddle-billed storks flying by while hippos slid into the water.

Champagne
Veggie pizza
Beef casserole with rice
Sautéed vegetables
Green salad
Garlic bread
Cheese plate

During siesta, we took wonderful outdoor showers, and then spent some time in the blind over the Chiawa River bed at the junction with the Zambezi. Delightful groups of baboon with babies, impala, bush buck, and warthogs cavorted right below us. Later, as we relaxed on the deck of our chalet, we watched two male buffaloes munching the grass right out in front of us, with the mighty Zambezi in the background. We wandered over to the main bar to get Mosis, and sat on the comfy chairs there, imbibing the ambiance of a vacation in Africa. Walking back to our chalet to get gear for our afternoon drive, a huge male elephant walked right in front of our deck, bold enough to advance toward the deck at one point before continuing his meandering along the grass in front of the row of huts. Yikes! Jim had to hurry, remembering that there's no running in Africa, or you'll be considered "prey," back to the main sitting room to pick up his solar panel that he'd left out on the grass. I joined him - a little more slowly - and then things got exciting. The elie came right up on the grassy slope to the fire circle where we eat our breakfasts and enjoy our pre-dinner drinks, selecting winter thorn seed pods with his trunk and throwing them into his mouth. Ellen and Ron were in the dining hut right in front of him, along with some other guests, and they had to be very careful about moving from that hut to the sitting room, so as not to irritate Mr. Elephant.

And for the next half hour, we were enchanted to watch this exceptionally large elephant (Joe guessed he was about 50 years old … perhaps 51 years and three days, like Jim?) as he made his way through the camp, finding his snacks along the trails. His eyes were very deep, and almost sad-looking. He never went into any of the enclosures, but came right up to the completely open sides and allowed us to get some photos of each of us with a completely wild elephant. I guess you could call this encounter #12 of the day…

Ellen and I needed a loo break before our drive so Barbs accompanied us as we walked from the bar to the little room. As we were walking quite near the elephant he moved directly towards us with a bit of irritation, which was quite scary. Luckily when we were ready to leave the loo hut, he had moved on. In fact, he'd moved right over to where the jeep was waiting, with Ron and Jim on board, so our guide Dispencer had to quickly drive the truck into the staff compound. Finally we were reunited when the elephant moved on down the road, and our game drive could begin.

"Bring lots of tissues for Cynthia!" Barbs called. Guess by now I was acquiring something of a reputation with all my crying over elephants. ☺

The moon was up and almost full as we enjoyed francolins taking a dust bath in the road, and then practiced our raptor identification skills. A large one was perched in a baobab tree, and as none of us could figure out what kind of bird he was, Dispencer drove closer and closer. Even when we were literally underneath him, he hadn't flown and we STILL couldn't figure out which bird he was. What irony: normally you don't want the bird to fly so you can get a good look, but here we wanted him to fly, so we could see his wings for a positive identification. Dispencer stepped out of the vehicle and immediately the bird flew off, so we could tell that he was a juvenile tawny eagle. Dispencer and Ron had a little fun with the next raptor in a tree; it turned out to be a bateleur but again we needed to get really close to be sure. Right after that sighting, we saw a fish eagle in a sausage tree. A night of eagles.

We spent time watching a mother baboon and her youngster, sitting in a tree in the most relaxed human-like fashion, with feet propped up. They each had a lot of red on their faces, so Ron dubbed them the famous Zambian red-faced baboons, even though we all knew it was from the flaming combretum they'd been consuming.

We had another elephant encounter, to make #13 for the day, out at Uys's Point on the river where we stopped for sundowners. Dispencer had to position the vehicle for a quick get-away in case they got closer, but they did keep their distance. It was a group of about eight elephants, with several babies who were entwining trunks and older females taking sand baths. Lovely.

As we left for our night drive, I was hoping to find a leopard so I asked Dispencer to find something with spots … "spotties" in his parlance … and he did! He found a large-spotted genet and a spotted hyena. Can't say he didn't deliver! I'd also asked for a porcupine and we did see one for a nanosecond, before it sashayed into the deep brush.

We then descended into the dry Chiawa river bed for a marvelous surprise, a complete braai (South Africa word for barbeque, pronounced "br-eye") in the bush. Light glowed from the lanterns strung all around, and the tables were set with white linen, china and stemware gleaming. Several bonfires were lit on the periphery as well as a huge fire over which the food was cooking. An elegant loo for the ladies was situated on the side - a toilet seat over the sand, with wash basin, surrounded by a cloth enclosure open to the stars. Lanterns hung from the cliffs behind. The most amazingly romantic setting ever! We were served drinks from the bar, and then after a bit of relaxing the Chiawa choir sang and … I cried. Craig described the exhaustive list of food arrayed around the fire in iron pots, from which we could serve ourselves. The dishes were chosen to honor both the settlers who had traveled inland from the African coast to find homes for their families, and the natives that they encountered.

Cooked whole pumpkin, filled with vegetables
Tomato/onion sauce
Greens cooked with onion
Tiny dried fishes (like anchovies) from Lake Kariba
Baked beans
Nshima (with cheese and tomatoes)
Beer bread
Jacket potatoes
Farmer's sausage
Chicken kebabs with dried apricot
Lamb cutlets
Bread rolls cooked on metate reeds
Couscous with vegetables
Cole slaw
Green salad
Well, that was enough for ME!

At the end of the meal, Craig got up and said that although he'd promised not to embarrass Jim, there were two guests whose birthdays had occurred earlier in the week, and that Chiawa would like to honor them. He asked Jim and Martin to come to the head of the table, and handed them a knife. "How are they supposed to SHARE a knife?" someone asked. Craig suggested that they hold it together, like at a wedding, and there was a lot of laughter about that, as they maneuvered their hands onto the large knife together. The staff appeared from the darkness, carefully carrying a beautifully decorated cake, topped with spluttering sparklers, while singing a Zambian happy birthday song. They placed the cake on the table and as the singing finished, Jim and Martin started to cut the cake. But they couldn't do it! They kept trying, but it was too difficult … the quite sharp knife simply wouldn't go through. Finally, after many attempts, the icing was smudged away from the cutting site and Jim leaned down, saying, "What IS this? It looks like grass …," only to discover that it was an iced piece of ELEPHANT DUNG!

Everyone broke into laughter, clapping as the staff came out again singing, and carrying a fantastically delicious birthday cake (chocolate with cream and strawberries). The last hurdle was the "magic" candles that wouldn't blow out, but as Jim has helped raise several teenagers, he knew how to douse them with fingers wet from a nearby person's water glass. What an honor, not everyone gets an elephant dung birthday cake. And so, in this memorable way, we had elie encounter #14 of the day.

During the night I heard a leopard cough quite near our tent, and baboon alarm calls. In the morning, there were many prints of the large leopard (the father of the one we'd seen on our first day at Chiawa) all along the camp path.

We awoke to the haunting scream of fish eagles and guffawing hippos. We had a virtual rain of mahogany leaves on our roof, as vervet monkeys were scampering about in the trees overhead. In fact, when getting our tea and coffee from the deck I'd left the lid of the box open on the table, and one enterprising monkey leaped down and grabbed the sugar dish. I ran to the front of the tent to discourage him and luckily this startled him so that he dropped it not far from the deck.

Our morning game drive started out auspiciously, as we received a radio call not five minutes from the camp informing us that the young leopard had been located on the ridge behind. We hurried off in that direction, and arrived in time to have a wonderful experience with him. He was reclining on the red sand, and essentially posed for us … turning his head in every direction so that we could admire it from all angles. After a respectable amount of viewing time, he then got up and walked RIGHT PAST the front of the jeep … stopping in front to pose for photos as he looked closely at us, and then continued on down into a gully in the brush. I decided to call him, "The Spirit of the Zambezi." It was very moving.

As we drove west to search for the lionesses again (remember, we hadn't found them on our previous lion hunt), I was enchanted to watch a large group of impala performing impressive graceful leaps through the air.

Dispencer related a story of a couple who had been guests in his jeep and had kept saying, "JAFI" every time they passed impala, and insisted that they didn't really want to waste their time looking at those antelope. At the end of their trip, he asked them if they would share with him what JAFI meant, and they happily explained, "Just Another Fucking Impala"! He got even with them later for being so insensitive as to not care about ALL animals in the bush, by telling them that he'd found a leopard, who was eating a kill, but he knew he shouldn't bother to take them there, as the kill was JAFI.

And suddenly we came across the lionesses. They were lying in a grassy dell, right alongside the tracks of the jeep road. In fact, one of the lions was lying across one of the tracks, and as they were all supine it looked like road kill! The oldest lioness, "Tag," was in the middle; she's 13 or 14 years old and we remembered watching her and her cubs last year. There was a young male lion, a sub-adult female (one of the cubs that we'd watched prancing and playing just last year), and two 7-month old cubs from this year's litter. About thirty yards away, sat Douglas, the large male of the pride. Their tummies were all quite full, and they lolled about, only mildly glancing in our direction periodically. It was peaceful in the grassy glade.

Later, after we'd driven some distance away looking for the kill that had been responsible for those full tummies, we saw that an elephant was wandering down the track in the direction of the lions. Dispencer suggested that we stop to watch. And WHAT an experience! As the enormous bull elie got close to the lions, they quickly stood up, including Douglas, even though they were quite full and had been lazily reclining. The elephant came a little closer and the lions were still standing at attention, so he decided it was time to send a message. He started to run towards the lions, with ears flapping and sand flying, trumpeting wildly. The lions scampered off into the bush.

So, who's the REAL king of the beasts?!

We saw vultures circling nearby, which helped our search for the kill remains. We found the vultures near a collection of bloody bones, skin, and hooves of a waterbuck. Several different species of vultures were hopping around on the ground or waiting in the trees overhead.

Along the riverbank after our stop for cokes and cookies, we watched a saddle-billed stork fishing and Dispencer gave us a story to help us to remember that the male has black eyes while the female has yellow eyes: the female punched the male, giving him a black eye, when he came home late after staying out and drinking at the local bar.

Back in camp we had a Mosi and a delicious lunch, then hot outdoor shower.

Beef or spicy veggie lasagna
Potato salad
Green salad
Tomato/onion bread
Cheese plate

Soon we were off for a canoe ride on a brilliantly clear and warm afternoon. We first motored up the Zambezi, with canoes in tow, passing slumbering hippos on the banks of the river and on the small islands, many with babies snuggled amongst their bulk. We unloaded at the "Grand Canyon" of the Zambezi, an area with deeply eroded sandy hills and gullies that really did remind us of Utah. After an excellent briefing by Daniel, including a paddling demonstration in the water, we were off. First we canoed down the Zambezi itself, then took off down the Chifungulu Channel for several miles before coming back out to the main river a few hours later. The Channel was charming, as it was slow moving and quiet and we could hear the myriad bee-eaters zipping around our heads, and the trumpeter hornbills crying. Brilliant white egrets dotted the grassy banks of the river, and we saw several gorgeous malachite kingfishers and immense purple heron. During the trip we managed to maneuver past a submerged hippo and a mammoth crocodile, thanks to Jim's expertise, as I paddled or back-paddled per his instructions.

We have canoed this channel twice before in previous years, but this time we actually got to stop and watch elephants cross. We'd never seen that before, and it was magical. There were five elies, including one quite small baby who got stuck in the mud and had to be assisted by another female pushing on his backside. When he was finally out of the mud he rolled about for a little bit, rubbing his bum against the ground, but then suddenly stood directly upright and defiantly flung his trunk at some egrets which had clearly been crowding him. He's going to be a tough one.

We later passed another group of six elephants who opted not to cross, so we watched them in the bushes on the bank as we passed by.

Back on the main river we could see the Chiawa camp ahead downstream. A table complete with white tablecloth and candle were set on the shore with our sundowners waiting. Grant was there to welcome us, and I cried once more. He'd planned to come to the camp with Lynsey and his newborn Scott earlier in the week, but Scott had fallen and they needed to make sure he was ok instead of flying out to the bush. They are hoping by next year to have their personal hut built, so that the whole family can be at the camp. Currently Lynsey and Scott are staying in Lusaka and Grant is commuting.

Grant then took us on our night drive, which was splendid. We stopped to quietly watch a family of elies by moonlight down by the river … sublime.

The mom was rumbling to the youngster, perhaps warning her to stay away from our jeep. We watched while they gave themselves a sand bath, and Grant explained that by throwing sand on their wet bodies and then rubbing against trees, the mud removal would pull parasites off their skin, like a good facial! The mom had a wound on her temple that was oozing pus, and Grant thought it might be from a bullet. He called on the radio to a vet who was visiting the park (trying to remove a snare from a hyena), to ask him to stop by later to check out this elephant.

As we were watching a genet in the crook of a baobab tree up on the ridge, we got a radio call that the young leopard had been sighted again. Grant asked us if we wanted to rush to see it, and we all agreed. We took off through the night, driving as quickly as possible through the bush. It was thrilling to have the cold air whizzing by, and Jim and I remembered that we'd had this exact same experience with Grant last year, and that the call had come in at the same spot on the ridge. Wow. We got to the field where Dispencer's jeep was parked, and saw the leopard. In a tree, he was relaxing on a large branch, EXACTLY as in all the photos in all the tour books. It was lovely watching him there, and I decided we should name him "Bracket," as he'd been the first animal we'd seen on our stay at Chiawa, and then the last animal we'd seen. The first time he was UNDER a tree branch and now, the last time, he was ON a tree branch.

Later, back at camp, we settled into the usual routine with drinks by the fire, the singing choir, superb dinner under the stars, and a night with lion calls and hippo bellows.

Shrimp cocktail
Rolls
Zucchini soup
Leg of lamb or bream with tomatoes and onions
Risotto
Snow peas with peppers
Pears with chocolate

In the morning we had our breakfast by the river, and then said goodbye to Ellen and Ron, who were off for their adventure on an island off of Mozambique. It was tearful to have them leave as I'd really enjoyed having them with us and sharing our love of Africa with them.

We spent the morning in camp, chatting with Grant for a few hours, looking at baby pictures of his son Scott and discussing how we can help with conservation work at CLZ. We then packed up our luggage and hung out in the main sitting room, chatting even longer with Grant and Craig, and ate a quick brunch before we left

Chicken or veggie curry, in a bowl made of roti
Rice
Green salad
Four-seed bread
Cheese plate

I wrote a short song for the guest book:

In the African bush, the magical African bush,
    the lion calls by night.
In the Chiawa tents, the happiest of safari travelers,
    the "clients" listen all night.
Elies in the moonlight, leopards in the trees,
    lunch cruises and candlelit braais.
We love you Chiawa, we'll miss you Chiawa,
    we'll be back very soon!

We hopped in the boat and of course I had a really good cry as we continued upriver to Chongwe Camp.

Just as I was telling Jim that I wished we could have stayed at Chiawa longer, we walked up the path to the main sitting area of Chongwe and saw ELIES EVERYWHERE!

Two large bull elephants were near the bar contentedly munching winter thorn pods; a mother and a baby strolled down by the tents along the river; a family of six with two tiny babies rustled in the hyacinth of the Chongwe River, right across from camp. And while we watched, a miraculous adventure ensued.

First, the mother and baby walked down to join us at the bar area. This is highly unusual - breeding herds of elephants with babies essentially never come into camps, as they are too nervous about their infants. But here was this mother elephant, happily scratching her bum against a tall winter thorn tree only yards from guests laden with cameras and videos, and tears streaming down our faces! She rumbled to her youngster quite a bit, especially as the child wandered closer to the humans. But this baby seemed so carefree, even with the mom's concern, literally almost skipping down the path, playing with the rocks lining the edges, sniffing at us, and scratching one leg against the other. Remarkable!

Eventually the pair moved off into the trees, and I walked over to the riverside to see what the larger family was up to. Just as I got there they started to enter the river, crossing to our side. Although the larger females were able to walk across, getting wet about three-quarters of the way up their sides, the smallest baby was so short that it had to swim, and was eventually completely submerged with only a tiny trunk poking up now and then like a snorkel! There was a large animal at each end of the train of elephants, protecting the smaller ones in the middle. They were rather crowded together, bumping into each other as they went, a protective strategy we've often seen. After they climbed up on the riverbank between two of the tents, the elies then moved to the wallowing pond that the camp had been building when we visited last year. I think this may have been the first breeding herd to play in it. They wallowed for a long time, and it was an exquisite sight. They trumpeted loudly, flailing trunks and flapping ears, flinging mud and pushing and shoving and rolling around in the pools. The babies were twirling their little trunks madly out of control, and stayed longer than the adults, rolling and flopping their whole bodies down in the mud. I think it was the most fantastic sight I've ever seen.

After a long time the elephants gathered together and walked off into the bush, and Jim and I were shown to our chalet. The tents are set on smooth concrete pads, with elegant concrete bathrooms on a lower lever behind. These bathrooms are the most open of any camp we'd been in, consisting of a simple thatch wall about six feet high with no roof overhead. As the trees were not too close to the structure, I was able to star gaze when visiting the loo at night. The concrete of the shower had stones set in it and a low wall of stones blocking the spray from the rest of the room, with the sink set into a wooden stand made with local tree branches. The tent itself was a bit small (not much storage space around the queen-sized bed), but wonderfully open to nature with huge windows all around. From the front of the tent we looked directly out on the Chongwe River, and watched the sunrise from bed in the mornings, while enjoying the sight of impala, hippo, and baboons on the far side of the river.

We had a delightful hot shower and relaxed in our tent for a while. Several bull elies wandered by the tent from time to time. One of them was badly injured and actually rested his head on the bathroom wall of the tent next to ours, damaging it completely so that it had to be replaced the next day. We had tea and snacks and enjoyed reacquainting with Silkie, the impala who has been at the camp for over ten years and whom we remembered fondly from last year. She had an arthritis problem, and the vet who was at the park this week (still trying to get close enough to the snared hyena to dart it) came by the previous day to help her by performing surgery on her ankle. He had to be quite careful with the anesthetic, however, as she was clearly quite pregnant!

Chongwe is located just outside the Park and so as we headed out for a game drive in the Park, we had to first pass through the entrance gates where

Levy, our guide, was well prepared with the necessary paperwork; we also had to exit the Park by 8pm. We passed a group of nine elephants drinking in the river, and watched vervet monkeys in a tree.

The moon was growing quite full as we came across the lion pride. We'd been hoping to see them again. Jim had thought he'd seen Douglas in a grassy gully, but it turned out to be a hippo skull. The lionesses and young lions were all in a circle dining upon a huge male warthog. Dispencer was already there with his jeep, and whispered to us that they'd actually witnessed the kill which included smothering the warthog for about thirty minutes before he stopped struggling, only moments before we arrived. We were glad to have missed that. Even still, it was pretty gruesome, although we did watch in horrid fascination for quite some time. The sounds of ripping skin and flesh and sinew, and the cracking of bone was unsettling, especially for a vegetarian! We were also surprised that the lions purred loudly and growled at each other the entire time that they were digging in to feed. Sometimes little fights would flare up over a choice morsel, but even without any actual contention they were still snarling and howling continuously. They'd managed to wrestle the warthog down in a muddy dambo, so several of them were covered in mud. Apparently the young male had tried to help out, but didn't quite get the concept as he came at the warthog from the wrong end and kept getting kicked away.

We drove off, a nice long way from the carnage, to exit the jeep and have our sundowners (well, sungoners again) and hoped that Douglas would come to join the warthog feast. However, even when we passed by them on the way home, he'd not arrived. He was "busy" with another pride of about fifteen lionesses further west in the Park with whom he was visiting. We saw one of the cubs grab the head of the warthog and succeed in pulling it away from the other lions so he could get his started on his meal.

On the drive home we enjoyed elies by moonlight at the edge of the river again, and saw a genet, a civet, and a white-tailed mongoose. And listened to the squeaks of nightjars and the flat battery call of dikkops.

Back in camp, we caught up with Lindsay and Garth over drinks, and the goings-on of the past year. They are now running the "Chongwe House" that was built last year, and really love it there. Lindsay offered to give us a tour the next day. We dined this evening down by the river.

Mushroom/potato pancakes
Rolls
Rice
Stewed veggies (with or without meat)
Okra
Nshima patties
Green salad
Cake

We happily enjoyed another fantastic night in the bush, listening to the plinking of tiny frogs and low hyena calls, and a hippo eating marshy grass right outside our tent all night. This whooshing sound was loud and wonderful, and the moonlight on the river awe-inspiring. I even managed to wake up at the right time to watch the moonset and admire Orion.

We slept late in the morning - we wanted a little "lie-in" as they call it here. Listening to geese fighting and a tremendous dawn chorus of birds brought us deep peace and joy.

While we were eating breakfast by the river, a yellow-billed kite swooped down low and grabbed a piece of toast from the pile by the fire, and then came back again - hitting Jim's head as he passed by to grab the butter dish! Guess he wanted his toast buttered.

Lindsey took us on the tour of the Chongwe House, and we were really impressed. The architecture is astounding – the house is built from stucco with extraordinary designs of local rocks and trees inset into the walls, with a towering thatch roof. There are four luxurious bedrooms and an extensive common sitting room, outdoor dining area, and swimming pool, right on the river. It's a perfect place for a family or group of friends to stay for safaris, and there is a dedicated guide, jeep, and boat for the exclusive use of the House guests.

Back in camp we watched a bull elephant charge and trumpet loudly at a new guest who didn't yet understand that he shouldn't walk out of the loo with an elephant standing near by. Quite scary. During lunch we chatted more about the heartrending fate of the famed Australian crocodile hunter, the only news we'd heard for the past three weeks (he'd been tragically killed by a stingray during filming of his TV show).

Chicken barbeque
Tomato stuffed with rice and curried tofu
Potato salad
Green salad
Onion bread

I actually took a nap at siesta time as I seemed to have caught a cold, and later, after our afternoon snack and tea we went for a boat ride on the Zambezi. Other than the put-putting of our vessel, the peacefulness out on the river was quite special. We enjoyed motoring close to the beautiful grassy islands filled with jacana, goliath herons, and plovers. Elies were wandering in some marshy grass, and hippos lolled about. We saw an Egyptian goose riding through some pretty stiff waves as the wind kicked up. The goose was followed by a single tiny chick valiantly trying to keep up.

As we sat quietly chatting to Ian, the other guest on the boat and a safari travel agent from Boston, we were wonderfully distracted by the rising moon, a full moon and quite red. Breathtaking!

Back at camp we were to hop in our jeep for a night drive, but had to tromp around in the bushes to reach the jeep, as there was a large bull elie in the pathway and we did want to avoid him. We had an eerie experience watching a crocodile gliding along a shallow section of river, with game in his mouth - an impala, or warthog, or perhaps a duiker. His two shining eyes looked particularly evil as he slowly sank out of view, taking his dinner with him except for one fuzzy ear still sticking up out of the otherwise empty water.

We saw the local Pel's fishing owl that we'd seen last year, perched quite nicely in a tree, and heard his call for the first time, a very poignant whiny whistle. He flew away on silent immense wings.

Another surprise was in store for us: dinner in the bush! Again, it was a lovely set-up, with an elegantly set table alongside the river, lanterns all around, and fires cooking the delicious food. And a cloth-enclosed loo in the trees!

Sweet corn soup
Rolls
Veggie or beef kabobs
Cheesy cauliflower
Potatoes
Profiteroles

While we were enjoying dinner, Jim pointed at the moon, saying "I thought you said it was a full moon tonight," as there was clearly a piece missing from the side. It was a partial eclipse! The ambience was perfect as we watched the dark part grow while dinner progressed.

We slept soundly, happily listening to thunderous hippo calls, strident geese fights, and melodic bird song in the morning.

For our last morning game drive, we enjoyed seeing two juvenile fish eagles on their nest, and about ten solo bull elies in a row. Guess all the ladies and children were sleeping in. We visited our "flat lion" pride again, who were clearly sated once more, strewn about on the grass as they were. One of the cubs and Douglas (who was now back in residence with this group) did eventually deign to raise their heads to give us the time of day. A breeding herd of elies including seven females and one tiny baby, meandered about; leaping impala were all around.

Herb bread
Pasta with veggies and tofu or meat
Eggplant with cheese
Potatoes with mustard
Green salad

Back in camp it was time for a Mosi and lunch, before siesta. We watched an adorable baby hippo, right across the river, walking around and yawning wide, before plopping back down next to his mother on the sand. While savoring my outdoor shower I heard an elephant approaching nearby. It was such fun to climb up on the wall of the loo and peer out at an elephant only a few feet away. We got some good photos of that event!

In the afternoon, Levy took the two of us on a relaxing bird safari in the local area. We'd seen 133 species of birds on this trip (10 life birds), and still wanted to watch for more. We enjoyed the sight of four three-banded plovers on the riverbank fighting with each other (such ferocious struggles from such tiny birds), while baboons climbed the red cliffs behind. We finally saw a red-eyed dove – we had been hearing them all along, but hadn't taken the time to see one. Along with the terrestrial bulbul in a tree, this made 135 species sighted, not bad for 17 days!

Relaxing on a grassy sand bar, we watched our last African sunset … and drank our Amarula out of cups made from a water bottle cut in half, as the dishware was misplaced from our cooler. We toasted each other, and … I cried. Three waterbuck posed for us, and we gave Levy Jim's binocular straps as a gift.

The moon rose, red again, and again, I cried. I will miss Africa.

Chicken or lentil casserole
Rolls
Couscous with veggies
String beans
Green salad
Butternut squash squares
Jell-O custard

Our last night was the noisiest yet, with a hippo munching the grass literally five feet from Jim's head and the geese insanely loud. The hippo slipped back into the water when I turned a flashlight on him to get a better look. It felt like a short night as we arose at 5:30am for the 20 minute drive to the Royal Airstrip where we quickly got into the Barron. Jim flew front seat again and in another quick 20 minutes we were once again at Lusaka for the usual British Airways flight at 8:45am.

Hours later, we were home. And I cried.

Chapter 5
Africa 2007

◆

"Crying in the beloved country"
or: "3 weeks with one pair of pants"

Within 10 days of returning home from our trip to Africa last year, we were already planning this year's trip. We knew we wanted to go back to the Lower Zambezi National Park (LZNP) to visit the owners of Chiawa, who have become good friends, and spend more time in their glorious safari camp ... as well as volunteer at the conservation group, Conservation Lower Zambezi (CLZ). And of course if we were in the LZNP we simply HAD to go back to Chiawa's "bush camp," Old Mondoro.

So we already had the outline of a trip.

Bushtracks, as always, helped us to take the outline and make it a reality. We added on a visit to Namibia as Jim had been so ill during much of our only visit to that country (apparently from eating too much meat with a stomach accustomed to my vegetarian lifestyle), and he really wanted to see the amazing red sand dunes at Sossusvlei while NOT throwing up. We also fancied a stop at Victoria Falls where we've only been once before.

We decided to go earlier in the season this year, as our son Mark was planning to get married in the fall. This meant that we might see fewer animals than later in the season, however we thought it would be interesting for us to get to see Africa at a different time.

After our itinerary was confirmed and we had secured flights using our frequent flyer miles, there was not much more to do except wait for July to arrive. We made the fatal mistake of watching some of Ron and Ellen's video footage from our trip together last year and their pictures were incredible. Ron had told us he'd purchased a "high definition" video camera but we had had no idea how remarkable the sharp picture would be.

Jim decided we had to get one too. And to tell the truth, I didn't object. This meant lots of studying and comparisons and evaluating and shopping, Jim's specialties. We were excited when he figured out the whole new system and got all the requisite components.

Another preparation for the trip was Jim's getting a tattoo. He normally doesn't approve of tattoos but he figured that since he's over 50, he can relax his strictures. And he really wanted a tattoo of our favorite animal … the elephant. He found a fantastic picture of a mother and baby elephant, standing close together, and he got this tattooed on the back of his right calf. He also had them add the phrase "panono, panono," which means "little by little," in the language of the local Zambian Bemba … the elephants are coming back from extinction … little by little.

We didn't get much of a chance to stray far from our connection to Africa when in the winter, Grant Cumings, the owner of Chiawa Camp, visited California and stayed in our home for a night. It was great fun to see him again, and we took him out for a wonderful meal with Ellen and Ron. During his visit I asked him again about whether he could purchase a set of carved wooden guinea fowl like the ones we've seen at Sausage Tree Camp (or steal them, if need be, ha ha!), as a secret gift for Jim. He promised to do the sleuth work necessary and when the camps opened in the spring he managed to take photos of the desired wooden guinea fowl and sent them to me in a series of clandestine emails, asking me to describe exactly what I'd like his woodcarving friends to produce (the size and number of animals). We agreed on a plan, and I awaited our arrival in the LZNP with great curiosity to see what they'd made for us.

We chatted with Ellen and Ron about our upcoming trip, and I became quite teary-eyed at the thought of the limo arriving on July 3 and our driving off to the airport without them. I knew that I would be very sad, but they were preoccupied with planning for their wedding in September and just couldn't manage the trip this year.

Packing for our trip began in earnest in April. Jim upgraded the solar panel and photo printing kit so that it fit in a smaller container, and painted a cute hippo on another small bag for our cameras.

We read the website from American Airlines to learn that we could each carry on one bag and one "personal item" (such as a purse or briefcase). Therefore, each of us could carry one duffle bag of clothes and one of the small camera/printer bags. I was quite proud of how lightly we packed! My duffle bag had my clothes (three shirts, one pair of pants, jacket, gloves, hat, long johns for sleeping, bathing suit, and undergarments), flashlight (aka in Africa as a "torch"), head lamp, journal, several books, birding book, binoculars, toiletries, medicines in a plastic tray, and flip flops. Jim had a similarly small set of clothes in his duffle but also several torches, green laser pointer for star gazing, all the camera charging adaptors, video tapes, tripods, camera clamp for safari vehicles, and a photo viewer for storing digital photos during the trip. We prefer to travel with only hand luggage as it eliminates the possibility of its getting lost.

I was also busy with the fun of organizing the paperwork, tips, and drugs. For staff tips and exit visas, I calculated the exact costs and how many $5, $10, and $20 bills would be required, made a trip to the bank for the precise denominations, and then put each tip and visa in little labeled envelopes (totaling almost $400) in my tiny airplane over-the-shoulder bag. Keep in mind that once we are out in the bush there are no banks, not to mention the fact that the US dollar goes a long way in the places we typically visit, so it's key to carry a lot of smaller denominations.

CLZ sent us an email in June mentioning that they had an orphan elie in camp … and we WISHED that it could stay at CLZ until we were there. They also reminded us of the donation we had promised to make, and Jim sold two of our three Krugerrands to help foot the bill.

I pulled out the advent calendar once again, and Jim suggested that this year instead of buying all sorts of little African animals and jewelry, we print out articles of interest about Africa to place in the pockets. It was a great idea, and each day we read to each other the articles we had found.

During the days prior to our travel, we had no small amount of worries about leaving. Our daughter Jessica was down in Honduras on a coral collecting trip, and emailed a sweet little "please call" message, to tell us that she was desperately sick, having contracted dengue fever. I couldn't believe it. So there I was, calling to shower her with loving motherly attention, while Jim was on the computer reading about dengue on the internet, calling out to me, "Tell her not to take any aspirin, only acetaminophen … oh, and if she starts bleeding from the eyeballs, tell her she should go to the hospital …"

She was really quite sick, with diarrhea, vomiting, high fever, and horrible headache for many days. Meanwhile, Mark had quit his job so he could take a two-week intensive scientific dive training class at Scripps and then work for Jessica for another two weeks down in Honduras, but he didn't receive his "guaranteed" passport delivery (said guarantee costing him a very hard-earned $150) in time, so that he had to miss his flight to Honduras and simply keep hoping that the passport would arrive.

That aside, I had frantic worries about all that needed to get done before we could leave, the work issues that might arise while we were gone, and the new project leadership mantle - the third project that I was now intimately involved in - having just been placed (dumped?) upon my shoulders. Further, the first injection dose of MY drug from Russia in a disease setting, which I've been working on for ten years, unfortunately had been delayed enough so that it would occur during our absence.

NONETHELESS, we were going to Africa, and so, we listened to the African sounds CD at night, drank Amarula, watched our previous year's videos, and dreamed.

Of Africa.

Meanwhile our beloved cat Billy became seriously ill, and the decision about what to do about his having lymphoma rather than the simple cold that we thought explained his symptoms had to be made literally five minutes before the limo arrived to pick us up. I'm sad to say that there was really no treatment option, so we had to say "goodbye" to him over the phone.

Diana arrived on our departure day, took our photos as she has done at the beginning of each of our Africa excursions, and shed some tears over Billy, whom she had tenderly cared for over the years. The limo arrived to whisk us off to SFO.

I felt torn as I was missing Ellen and Ron, frantically sad about Billy and calling friends and family to inform them of his being put to sleep, and calling Mark to wish him "happy 21st birthday." One of our more emotional departures.

Of course I cried.

When we arrived at the airport and got in line to get our boarding passes, I didn't quite understand why two young girls were running towards us and laughing. It was the two project managers who report to me, Cindy and Patty, but I didn't recognize them out of context! They had arrived to say goodbye, and to give me a beautiful gift of fun photos and cards, carefully labeled for each day of the trip, so that I wouldn't forget my dear friends at work.

Wow. I cried again! It was an incredibly thoughtful gift which had required a lot of work. I felt really quite honored.

Finally we left. First we flew from SFO to LA; as we were using frequent flyer miles, the route to Africa was rather circuitous. In LA we opted to go back out through security to ask for bulk-head seats for the flight to London but no luck, they'd already been taken; however it also meant that we had to go through security - again.

And what a fiasco!

The TSA employees there decided that our painstakingly specially purchased and prepared 3 oz containers didn't pass muster, as they weren't STAMPED as being 3 oz. Mind you, they were SMALLER than the stamped and acceptable 3.3 oz container of contact lens solution, but the TSA employees argued that they "couldn't tell" that the one was smaller than the other. So they confiscated all my Listerine and sunscreen bottles! Jim and I both threw fits, naturally, marching up the chain of command and insisting that we be allowed to carry these items on board with us. But, we really have no power. At one point I said I was NOT going to leave without my mouthwash to which they responded, "Ok, don't go on your trip, fine with us!" Hmmm. We were so irate that they finally called American Airlines, who sent an employee to us to box up the six offending bottles so they could go as checked baggage. I received the baggage claim ticket, but still have never received the baggage. Wonder where that box of little bottles ended up? Last we heard, they were somewhere in Namibia.

We flew from LA to Heathrow, London, arriving on July 4. Somehow it seemed that they were still blaming us personally for 1776. Now, while I don't want to bore you with details of our travel woes, this information may be useful to YOU one day. Although we had carefully checked the American Airlines website, we had neglected to check the British Airlines (BA) website - the airline that was to take us from Heathrow to Johannesburg (usually referred to as "Joberg"), in South Africa. So, we didn't know that we could only have one carry on bag and the rule was strictly enforced. If you wanted to carry a book, for example, that would be your one carry on item. So we had to scramble at the last minute, deciding how to deal with our luggage. It goes without saying that if we had known we'd have to check a bag, we would have opted to bring a larger camera bag, with enough space for a change of clothes and all the "essential" items, like camera adaptors and medicines, and so on. We were rushed to grab things out of our duffle bags so weren't able to be very clever. We couldn't fit anything into the small hard-sided camera bags, and yet we weren't allowed to carry anything in our hands! I managed to squeeze the contact lens solutions (that we use on the airplane) in one pants pocket, and a book and all our paperwork in the other, and then I pulled out my binoculars and hung them around my neck (funny, they were considered "clothing" since they were attached to me). But the little bag with all our tips, which I shoved into the duffle bag, and our medicines and toiletries and clothing and journal and bird book had to be checked. British Airlines kindly checked the bags all the way through to Windhoek, Namibia.

Before our flight, however, we still managed to conduct our special ritual: playing travel scrabble with our vodka drinks and chocolate. It just wouldn't seem like Heathrow without that!

After our five hour layover, we flew the next 11 hour flight from Heathrow to Joberg fairly comfortably. We find sleeping pills tremendously useful on these long legs. In Joberg we had another few hours to wait, and then were quite happy to be upgraded to first class for the two hour flight to

Windhoek. We joked about why we'd been upgraded: because they felt guilty about losing our luggage perhaps? Now there's a joke …

However, when we landed at Windhoek, our luggage was indeed missing! Even after all of our cautionary measures. I was proud of how kind we were to the harassed BA employee at the airport taking down all the details on our bags - literally half of the people on the flight didn't receive their bags. We then met our pilot who was to fly us to the Namib desert in southwestern Namibia.

The two hour flight in a tiny Cessna 210 would have been more fun if we'd had earplugs (they were in my bag) or if Jim had his Dramamine (that was in his bag). But as we taxied out for takeoff, we saw our first birds in Africa: guinea fowl running across the runway. We flew fairly low the whole time, the pilot gauging the green-ness of Jim's face during turbulence, to decide whether to increase the altitude to less bumpy air. Jim had the airsickness bag CLUTCHED in his hands the whole flight but just managed to avoid having to use it. Yeah, quite an accomplishment for Jim – his first time flying in Namibia without vomiting.

As we flew in to the little dirt strip in the NamibRand Preserve, we were equally overwhelmed and enchanted by the views. The desert was flat and covered with yellow Bushmen's grass, while also striated with brilliant ribbons of tall red sand dunes spotted with clumps of green grass, and pockmarked with myriads of sandy fairy circles. In this area there were literally thousands of these mysterious-looking circles – devoid of vegetation – ranging from two to five meters in diameter. After landing, we were warmly welcomed by Petrus, our driver and guide for the next three days. We hopped in our jeep and drove the short distance to the Wolwedans reception complex, housed in the buildings of the former ranch that had occupied this land. As we drove, we were captivated by seeing the elegant and delicate springbok running and leaping, the ground squirrels standing to gaze at us with hands hanging in front like meerkats, and two very large birds striding through the grass: Ludwig's bustards, the second heaviest flying birds in Africa.

In the stone tiled courtyard of the reception buildings we listened to the musical chattering of the black-eyed bulbuls in the trees, sounding like "Quick! Three beers! Quick, quick! Three beers!" We gazed up at a spotted eagle owl, aloof on his branch but following us with alert eyes.

We explained that we had arrived with literally no luggage; no jackets (it gets very cold in the desert in the middle of winter), no toothbrushes, no comb … nothing. They had a little gift shop and so we picked out matching fleece jackets - truly matching this time as the only size in stock was large, toothbrushes and toothpaste, and a tube of children's sunscreen. One of the women in reception loaned me a comb.

We then had a fifteen minute jeep drive to the Wolwedans Lodge, but which took longer as we kept having to stop for wildlife sightings: yellow mongooses popping in and out of their burrows, streaking along the ground in between the various entrance holes; a bat-eared fox stopping to stare at us with humongous ears and bushy tail; ostriches daintily meandering; and several local birds. My favorite was the pale chanting goshawk – affectionately referred to as PCG – which is a large pale gray raptor with red legs, long black tail, and bright pink beak with a black tip.

Petrus drove us around to the row of six chalets, built upon wooden decks along a ridge of the red desert sand. We climbed up to step inside one, and I was completely speechless. All I could do was cry! The chalet was essentially a wood and canvas backdrop to the wide-open view of the intensely red desert sand dunes, sparsely covered with clumps of luminous green and yellow grasses waving in the slight breeze, with the towering backdrop of the stark purple Losberg Mountain rising in the distance. There was a sumptuous double bed with down quilts, thick pillows, and mosquito netting draped above (for romantic touch only as we never really needed any netting in our trips to Africa, except one time when we had a bat join us during the night); a smooth wooden deck out front with lounge chairs for gazing at sand or stars; and a small wooden bathroom with shower, twin sinks, and large open windows looking out to the wilderness.

Cape and great sparrows were flitting around in the eaves over the deck and even came right into our room. After deeply appreciated showers, we put the same clothes back on (a little yucky after all that travel), and walked along a wooden boardwalk to the main lodge. And this was even more beautiful! There were several large rooms, all arranged in a semi-circle with extended outdoor decks, facing towards the mountains and a small waterhole in the grassy red sand out front. Some of the rooms were for relaxing, with deep armchairs and filled bookshelves; others had tables for meals and a bar sparkling with bottles.

By now it was dark and there was a fire burning in the fire circle in the deck. We enjoyed our first Amarula and met Peter Bridgeford and his wife. Peter has written, "Touring in the Sossusvlei and Sesreim," and the couple was visiting the camp in honor of the following day being the opening day of the new research station that Wolwedans has built on the reserve. Quite erudite company! Francis, the host, showed us down to the well-stocked wine cellar where a local wine was being sold to raise money for an orphanage/daycare for staff on the property. It was most enjoyable and really so nice to relax in this way after our long trip. The delicious dinner menu was announced by Francis in English and by the chef in the local click language, Damara-Nama. Namibia was formerly part of South Africa and therefore people who grow up in the larger cities speak Afrikaans and English whereas the desert dwellers from small native villages speak one of several dialects of the Nama (previously known as Hottentot) language. This language is a national language in Namibia, and we loved listening to the frequent clicking sounds which are sprinkled throughout.

Papaya soup
Veggie stack
Bream with vegetables
Cake

Walking back to our chalet using the torch loaned to us (remember, ours were "in our bags," an expression we really started to use in earnest), we delighted in the vast dome of dazzling stars. The Southern Cross, Jupiter shining amongst the stars of Scorpio, Venus setting near the horizon.

And absolute silence.

During the night, we heard one lone cape fox yelping. The half moon rose at midnight.

We'd asked for a 5:30am wake up call with a thermos of hot water, so that just as Orion was rising we could see – from the warmth of our bed – the sublime orange/red glow over the mountains as the sky began to lighten. Jim jumped up to make our coffee and tea, then climbed back under the covers so we could watch the arrival of the sun, streaking over the red mountains and the red sand dunes.

Walking over to breakfast, we heard a sound that was familiar from watching David Attenborough's "Life of Birds" DVDs. Large flocks of burbling, twittering Namaqua sand grouse were flitting to and from the water hole. We had a full breakfast on the open deck, watching several species of birds landing on the railings and hopping around the decorative clay pots planted with grasses and succulents. I asked the hostess why the thermos of water had been taken from our room the night before. I was parched during the night and really wanted to drink, and she explained that they had taken it to use for the hot water in the morning. "But I need to drink water during the night!" I exclaimed. She said that I should just drink the tap water as it was filtered and UV irradiated, that's why they had provided glasses next to the sinks, which was reassuring.

I then asked for a small container of vodka, so I could have some semblance to a Listerine mouth rinse, a little gross early in the morning I know, but it does the trick!

Petrus arrived to take us on our first game drive and what a lovely morning we had. Driving south from the Lodge, we wound our way through numerous sinewy red dunes. I simply could not tire of the beauty of the clean, soft, red, RED sand, and the incredibly incandescent green waving grasses. Petrus taught us all about the desert; the ecology of the various kinds of grasses, the various beetles and trees. He expounded on the fascinating theories about the fairy circles, i.e., termites versus a toxic latex from a theoretical past growth of euphorbia. No plants will grow in the circle and yet when sand from the circle is taken away and planted in a bucket, plants WILL grow. The mystery is still not solved.

Jim decided that the Ruppel's koorhan look like plovers when standing, like geese when flying, and like cranes when landing, what a combination! Their large multicolored wings were quite beautiful when outspread.

We watched a tiny steenbok, with sweet little horns, and then came across several groups of gemsbok oryx, gorgeous stately animals about the size and shape of a horse, but with striking black and white markings on their faces and a pair of exceptionally long, thin, impressively straight, swept-back horns. We came across many birds (our count was already at 22 species, four of which were life birds), and I thrilled to watch the ostriches running across the desert pan, their stupendous bustles jiggling. It was hard work birding in the bright sunshine without my missing sunglasses but still, I loved seeing many bat-eared fox, mongooses, and ground squirrels.

Back at the Lodge we had a romantic three-course lunch for two out on the deck, complete with white table cloth, china, and Windhoek lager, the local beer. Gemsbok, springbok, and bat-eared foxes came to drink at the waterhole, while sparrows and familiar chats made themselves quite familiar with the bread and butter on our table.

After lunch we delved into our usual staff photography session and set up the equipment to print photos for everyone. They all gathered round and were quite thrilled but the printer wouldn't work. Jim tried many times to fix it with no luck, of course the one piece he knew he could have used for repair was – yes, you guessed it – in his bag. We promised to mail the photos when we got home.

Petrus drove us to our next lodging at the Wolwedans "Dune Camp," a short fifteen minute drive away. The Dune Camp is more rustic than the Lodge, more of a traditional bush camp. Once again Petrus drove us up to our room, and once again I cried at the sheer beauty.

This room was an actual safari tent, erected on a platform of soft wooden slats, with the deck reaching far out front towards the sand dunes and mountains. Inside we again had a sumptuous double bed with clean white down-filled linens, but it was a much smaller, cozier space than the chalet at the Lodge. The loo was not attached but accessed via walking across a wooden boardwalk where there was a loo, double sinks, shower, and large open windows in the canvas, for looking out at the sandy wildness. We had the furthest tent in the group of six, and so from our room we looked out in all directions to the wilderness. The tent was surrounded by undulating red sand dunes - we could literally step off the deck and go hiking amongst the sand and grasses. And we did! We felt like explorers – or kids – as we raced each other up to the tops of the dunes and took photos of each other. We wanted to get a shot of the two of us together but with our tripods back in the lost bag, we instead carefully set up a photo shot where we each pretended to hold the hand of the other, and then we'd stitch the photos together when we got home.

The shower was luxurious, and the water drained out to a small acacia tree planted next to the walkway. Must be careful with water in the desert! We were ecstatic to have a chance to wash our undergarments and socks, and I held them in place on the deck railings in the sun with rocks as it was a bit breezy. Jim took a nap while I exercised and daydreamed, all the time gazing out at the dunes. I couldn't read … didn't have a book … they were also in one of "the" bags. Oh well, I never tired of the view.

Petrus picked us up later in the day and drove us to a giant red sand dune nearby for sundowners, adorned with table, tablecloth, candles, Amarula, and snacks. We climbed the soaring dune, which actually wasn't very easy as our feet sank so deeply into the very soft sand, and took photos from the top. I worried about our footprints damaging the sand and Petrus laughed. He assured me that by tomorrow you wouldn't notice anyone had been there. The wind sculpts these dunes.

I also have to admit that I was ever so slightly thrilled to learn that Brad and Angelina stayed at this Dune Camp when they were in Namibia, not just the thought of their celebrity, but the fact that teams of people were no doubt involved in finding them the best place to stay and I'd come up with the very same place all on my own! Well, to be truthful, not on my own as our camp hosts from a previous trip years earlier had told us that Wolwedans was a great place to visit.

Back at camp, we had drinks around the outdoor fire pit, enjoyed star gazing, and then moved inside the main tent, or lapa, to the elegant long candle-lit table for dinner. Again we had the menu described in English by our host James and in Damara-Nama by the chef, Martin. Earlier, Martin had been very solicitous about my dietary needs, pulling me aside for a private conversation to ensure that each course was acceptable to me.

Game meat/avocado salad
Home-made taglietelle with (or without) prawns
Brussels sprouts
Beef (or stuffed tomato) on sweet potato puree with broccoli and cauliflower
Cream custard in delicate orange candy basket

We had a good time chatting with the other couple at dinner, Tom and Cheryl, who were originally from the US (Texas and California) but now lived in South Africa building safari lodges.

We walked to our tent in the silent darkness of the desert, reveling in the deep sand underfoot and the immensity of stars overhead. We climbed under the fluffy down comforter and discovered hot water bottles. Perfect!

It became quite windy, however, and with that, cold. We decided to close some of the side flaps of the tent. I suggested that Jim get up to do the job; Jim suggested that I do it. Since I had been the one who had opened all the flaps earlier in the day, insisting that only sissies needed the flaps closed, I decided that I'd better simply get up and do it. So I did. And when I finally leaped back in bed, freezing to death, I slowly slid my hands over towards Jim … planning to warm them up at his expense.

During my trips to the loo in the night, I could gaze at the stars as I walked along the connecting boardwalk. At one point the Southern Cross was right over our tent; later the half moon rose over the mountains, looking like a bowl; and even later the moon was overhead and it seemed as bright as daylight.

Morning arrived with another gorgeous sunrise viewed from the comfort of our bed. This time the orange/red glow was sweetly framed in the "A" of the tent opening. The sand was a delicate pinky-red in the half-light. Our hot water was delivered and it was my turn to make the coffee and tea, and hop back under the covers. I couldn't stay for long, however, as I simply had to take photos of the striking views and our wonderful surroundings. I was particularly enchanted with the rocks up on our wooden railings, no longer holding wet undergarments in place but still looking austerely beautiful against the backdrop of the sand and mountains.

Breakfast in the lapa was delicious, and I liked that the back flap of the dining tent was wide open, so we gazed right into the soft red sand of the great dune behind. Petrus arrived to take us on our full-day game drive, but before we left I asked if the reception had a satellite phone that I could use. We stopped by the reception buildings, and indeed there was a phone. I called the Bushtracks office in Livingston (our next destination) to ask them if they could arrange for a few things at our next camp. She was quite sweet about taking notes for my requests: malarone (anti-malaria medicine); camera battery charging adaptors (our batteries were running down), video tapes (we only had the one in the camera), and some clothing and mouthwash. Fortunately, the hostess Zoe loaned me a pair of sunglasses for the day, which I greatly appreciated.

This day was filled with wildlife and beauty, as we drove north from the camp deep into the NamibRand Reserve. We drove along the expansive desert pans of short bushman's grass, filled with fairy circles, and surrounded by mountains on each side. We could see herds of mountain zebra and red hartebeest dotting the hills. Towards the northern end of the long valley we angled up into the foothills, where there were more birds and even a waterfall in the wet season.

Our first encounter was a family of gemsbok, a male and female, with a tiny baby no more than a few weeks old. They ran across the road and out into the grassy veld, even the baby! Then we watched several majestically large raptors – black-breasted snake eagles – that seemed to follow us along our drive, alternately soaring and perching.

And then, we were all alone in the wide expanse of the desert pan. Alone, that is, except for the posing ground squirrels, racing mongooses, and skipping bat-eared foxes. We came to a watering hole left by a previous rancher where gemsbok were congregating in large numbers, along with a black-backed jackal. The gemsbok were testy, as we watched several skirmishes and even one serious fight between two bulky males. It was quite interesting to see them charge each other, with heads tucked down far enough that they would strike their horns against each other. At one point the tussle even ended with one of the males knocked to the ground.

As we turned from the desert floor to drive along the trail leading up the sandy alluvial fan from the mountains, we encountered more and more birds (chats, larks, and weavers) and Petrus showed us the strange local euphorbia (woody bushes or trees with a caustic, poisonous milky latex sap), called the quiver tree that looked really bizarre, as if one took a small potted succulent and allowed it to grow 15 feet high.

And then we saw five giraffe! Two females and a male slowly marched amongst the rocks and sand, eating from the taller trees, along with two youngsters of about eight months old. The adults had been introduced to the reserve the previous year in an effort to coax the reserve back to the way it had been before ranching chased out the wild animals. Already each of the females had given birth, a good sign of a healthy ecosystem.

For us, however, our favorite animals were the springbok. We loved to watch them pronking. This is a delightful activity in which the male antelopes show off their prowess to each other, or to potential predators. It involves their leaping directly up from the ground, with their legs completely straight and pointed downwards, and their necks and heads similarly cast down. Boing, boing, boing, they leap and leap, quite high, fluffing up the bright white furry patch on their rumps to add to the spectacle. Absolutely captivating.

At one point Petrus suggested that Jim and I take a little walk along the trail while he went ahead with the jeep to set up our lunch. This was fun, as it's not often that we get to walk about, unaccompanied, in the African bush. The guides know SO much more about what to do if you inadvertently encounter a lion, leopard, or buffalo. When we joined Petrus under a spreading tree, he'd set up a complete feast. Picnic table and chairs, tablecloth, china, tea and coffee, three-course meal, and cookies to further sweeten the deal. As we sat enjoying the repast, plenty of small birds hopped about, and we even stimulated the attentions of a bokmakerie, a stunning bird with bright yellow and black front, gray head, and brilliant green back. Nearby in the rocks of the hillside we heard a strange repeated keening sound, with a little laugh at the end of each moan. We used our binoculars to discover that it was a small rock dassie, or hyrax, which looks like a small guinea pig but is the closest relative to elephants, surprisingly enough. I found it very endearing to watch him open his tiny mouth and have this enormous sound come out.

On the drive home through the desert valley we eagerly watched for more pronking springbok, running gemsbok, towering cliffs of rocks, and sandy pans of fairy circles.

To make a fabulous day even more wonderful, we arrived back at camp to the superb vision of my bag on the steppe. There was still no sign of Jim's bag (or our box of mouthwash bottles I might add) but at least we had ONE bag. The one with our anti-malaria meds, whew! I was overjoyed to have my bird book in which I've written all our sightings for the past five years, the jacket/hat/gloves which we've worn (matching of course) for the past five years, and the photo gift from my employees.

However, I soon realized that although the duffle was still locked, with the TSA-approved lock no less, it had been vandalized. My little over-the-shoulder bag, on which Jim had ironed-on a little lion, was gone along with all our tip and visa money. I felt sick at being violated and at having been so stupid at Heathrow to have forgotten about that money being in the little bag. It would have been different if the money had gone to people who really needed it, but apparently there is a large crime ring at Joberg airport and this was probably the efforts of just a group of criminals who'd gone through my luggage. I also lost my very nice torch, and our precious travel scrabble. How COULD they?! What a strange thing to loose. Anyway, with Jim's bag not arriving and the fact that it had had a lot of expensive equipment, we were resigned to the fact that his bag may never show up, along with his matching jacket/gloves/hat.☹ Darn. It was particularly irritating to have been robbed when we actually come to this continent to try to make a difference, to help the people in the small villages so that they can have living wages. We believe we contribute so much, and so to have it yanked away made us feel especially violated.

I changed into some clean clothes, what a treat, and we had our Amarula sundowners while relaxing by the fire circle. More star gazing as night fell, and then another unbelievably delicious dinner in the lapa.

> *Veggie spring roll over coleslaw*
> *Butternut squash soup with curry cream*
> *Springbok or veggies wrapped in lettuce*
> *Roasted potatoes and zucchini*
> *Chocolate mousse*
> *Lemon anniversary cake*
> *(it was Tom and Cheryl's 30 year wedding anniversary)*

Another silent night, filled with stars. I went to our loo to get ready for bed, and realized that the container of vodka which I'd brought to the room earlier to use as mouthwash had been thrown away. I was frustrated at this as I was feeling the onset of canker sores. I took the torch and walked back along the sandy trail to the main lapa to get some more vodka. It was a little embarrassing to have to ask for a glass of vodka after all the Amarula and wine we'd consumed at dinner, and it was a little scary walking in the dark by myself, all of which was adding to my irritation. My chapped lips were also bothering me but of course our Neosporin, which works so much better than chapstick, was also in the missing bag. I returned to the loo and finished getting ready for bed, walked across the walkway to our tent, and hopped in bed.

And couldn't find my hot water bottle.

And about then I lost it! I started to cry and, possibly for the first time in Africa, it was not a happy cry. I thought of all the stress I'd been under: work being so hard, Jessica being sick, Mark not getting his passport, Billy dying, Adam being robbed last month in Australia, Jessica and Mark being robbed last week, Mark's wife Stephanie's car being stolen right before we left, my luggage being vandalized, Jim's luggage missing, my canker sores starting, my vodka gone missing … and now they forgot my hot water bottle!

I felt as if I were going through the Kubler-Ross stages of grief: shock, irritation, persecution, anxiety, hope, despair, anger, dread, regret.

And Jim handed over my hot water bottle.

He'd simply been teasing me by grabbing them both. He was rather surprised at the strength of my reaction.

I apologized for being such a weakling, and picked up my book to read (I now had my books again). The book happened to be, "My Life in Rwanda." Talk about being embarrassed for sweating the small stuff. I ended up laughing at myself, crying over things so inconsequential compared to the real tragedies that happen in this world.

And so I moved on to the next stage: acceptance. And we had another lovely night's sleep, in the beauty of the silent desert.

The next morning we were up early to see the sunrise. After a quick breakfast, we had to check out, say goodbye, and hop in the waiting airplane at the dirt strip. The flight to Windhoek was pleasant at that time of day so no barf bags needed. When we arrived we made another check with the fellow at the BA baggage counter but no luck on the missing luggage.

We went though security to get on out next plane, laughing at the comparison with our experiences at LA and Heathrow. First, I didn't bother taking off my vest or binoculars, so the equipment beeped quite loudly as I walked through. The gentleman sitting there just smiled at me. Then, when Jim's bag came through the guy looking at the screen asked him, "Do you have a big knife in there?"

"No!" Jim exclaimed.

"Well, do you have a tube of toothpaste, then?" the screener asked.

"Why, yes," Jim replied.

"Oh, ok, then," was the jovial answer, as Jim calmly picked up the uninspected bag and walked out to the plane.

We flew in this small plane for a few hours, first stopping in Maun, Botswana to unload a few passengers, and then landing in the town of Victoria Falls, Zimbabwe. This was our first time to Zimbabwe. In fact, since we then immediately drove across the bridge over the Zambezi River into Zambia, we were in four different countries on that one day: Namibia, Botswana, Zimbabwe, and Zambia. We felt quite worldly!

Going through customs, we were excited to get Zimbabwe visas stamped in our passports, but were unhappy to find out that it cost us each $30. Oops, we didn't know about that expense and by now our funds were quite limited having had most of our US cash stolen and no ATMs from which to get more US cash. But, as we came out through the door of customs, firmly clutching our remaining baggage, we were overjoyed to see our sign:

TUTHILL X 2

Such fantastic memories! We laughed and followed the fellow holding the sign to his little van just like the last time we were arriving in Zambia. We drove to the Zimbabwean town of Victoria Falls and stopped at a few shops to see if we could get clothes for Jim, or video tapes, or camera adaptors. We were out of luck on all accounts since it was Sunday and most shops were closed, not to mention there were only a few shops to choose between. Even the stores with regular clothing were closed and Jim didn't want safari logo leisure wear, the only items at the tourist shops. A second driver then drove us across the bridge to Zambia, which took only a few minutes. He helped us clear customs on the Zimbabwe side ($30 was an expensive visa for a 15 minute stay), and then helped us go through customs on the Zambia side.

I broke into a nervous sweat as we stood in that line, remembering all the other times we'd flown into Zambia. The Cumings family, owners of Chiawa Camp where we'd be later in the trip, usually met us at customs and arranged for our visa waivers. Without the waiver, the visa fee is US$100 each. Bushtracks had warned us about that, and so we had been prepared that sometimes they don't waive the visa. I didn't think we actually HAD $200 remaining in our possession, and so I was quite concerned. But after some wrangling between our driver and the folks behind the windows, we were handed our passports, free to enter Zambia with the fee obviously waived. Phew! I did seem to notice at one point that he told them we were staying for three days. I knew we were going to be in the Victoria Falls area for three days, so I thought perhaps that's what he meant. You can see where this is going perhaps, but we didn't. I'll let you know how that little experience unfolded later.

Anyway, at this time we felt quite jubilant at finally being in Zambia, and loved seeing the quick view of the glorious Victoria Falls - "the smoke that thunders" - as we crossed over the bridge. We suddenly recalled that this town of Livingston had cell towers, and Jim discovered that his new iPhone worked. We called Bushtracks to see if they'd had any luck finding our camera adaptors, memory cards, video tapes, or clothes. They had not found anything, unfortunately, and all the stores on this side of the river were also closed. We stopped by the two large hotels, the Zambezi Sun and the Royal Livingston (where we'd stayed at when we'd been here years ago), and tried the activities center at each of those hotels. The answer was the same all around: normally they have video tapes and memory cards but they were just out of them today.

We continued the drive to our next safari camp on the river, about an hour's drive north from the falls, called the "Islands of Siankaba." As we arrived at the reception building, we could smell the enchanting mashed-potato perfume of the potato bush. We were finally back in the Zambia we love.

The safari camp was located on two adjacent islands, on the Zambezi River. The reception building was on the "mainland," and so we were shuttled by boat launch to the main lodge on the larger island. The boat skirted around this larger island, which was about a quarter mile long, to arrive at the side facing the expanse of the Zambezi. The boat dock led up wooden steps to the lodge which consisted of two elegant and spacious rooms, completely open on the side facing the river. The cool tile floors, comfy couches, full bookcases and complete bar enticed us to relax as we chatted with Georgie, the hostess.

Facing out from the lodge rooms, the view stretches across a multi-level dark wooden deck down to the water of the river, overhung with waving grasses and weaver's nests. The foliage around the lodge rooms was incredibly lush – extraordinarily tall thick mahogany trees with draping lianas and branches crossing every which way. It really felt as if we were in the jungle, surrounded by deep, thick, enveloping trees while the silvery expanse of the wide Zambezi allowed for just the right amount of open sky. The river was apparently the highest it's been in 40 years, so we reveled in the sound of the nearby waterfalls and a chorus of crickets. Unfortunately we didn't hear many hippos (I thought to myself, "But, I was promised there'd be hippo sounds…"), because they don't like such deep water. Apparently we'd experience the hippos if we'd come later in the season. After a refreshing Mosi in the lodge, our first Mosi beers of the trip, Georgie took us to our chalet.

The chalets were on a different island, reached via swinging suspension bridges over the small channel between the islands, amusing beyond belief. It was as if we were playing on Tom Sawyer's island at Disneyland. The bridges have quite a bit of sway to them, lots of bounce, and so Jim never tired of waiting until I was right in the middle before shaking the rope sides or jumping up and down. Apparently all the guys did this as you often heard giggles and shrieks from the ladies. There were seven elevated tree house chalets in a row, quite well separated from each other and designed so that from each room you could only see the river RIGHT out in front of the spacious deck of each room. Towering trees and vines were all around. The chalets were quite decadent: smooth dark wood floors and canvas walls, but the similarity to a safari tent ended there. The lofty ceilings were cathedral height, and the front tent opening was thereby immense. The king sized bed, filled with white fluffy comforters and pillows, mosquito netting above, was on a level with the open deck. On that level there was also a large writing desk, tall armoire, and deep plush chairs. Behind the bed was an upper level to the room for the bathroom where along with a shower was an immense claw-foot tub for two, twin sinks, fancy aromatherapy lotions and potions scattered about, luxurious towels and bathrobes, and elegant thick pile oriental rugs. Not only was the room so large and well-appointed, but there was also electricity. Instead of hot water bottles were electric mattress pads; instead of a carafe of cool water they had a small refrigerator with bottled water. Of course I did not want to drink bottled water, refrigerated no less, out in the bush. Luckily there were drinking glasses next to the sinks from where I poured my water to drink during the night. The fact that there was an actual hair dryer in one of the drawers was also an insult to our wilderness sensitivities, until Jim realized that this way he could wash his only pair of underpants and then actually dry them even if we didn't have time to hang them out in the sun during siesta. Cool!

We immediately took a splendid hot bubble bath, and Jim washed my hair. He knows how to endear himself to his woman, doesn't he?

Georgie had been able to loan us a charging cable for our video camera, and even gave us a second video tape. We were overjoyed.

I then realized that when we left this town in a few days we'd be flying into Lusaka to transfer to our small plane taking us out to the bush of the LZNP. And that meant that we could possibly meet up with Jenny or Dave,

Grant's parents who help run Chiawa with him, and who live in Lusaka. As Jim's iPhone worked, we called Jenny right from our chalet. I asked her if she could buy some items for us, Jim explained what we needed, and she was, as always, incredibly helpful, saying she would be off to the stores soon.

When it got dark we wandered back to the main lodge area, crossing the fun bridges and play acting as if we were on the "Pirates of the Caribbean" ride. Giggling in the dark, surrounded by giant dark trees draped with thick vines, tiptoeing on raised wooden walkways elegantly lit by kerosene lanterns reflecting on the water below. A roaring fire in the fire pit welcomed us and we settled around, drinking our Amarula and chatting with the other guests.

Dinner was served on long tables out on the deck, under the stars and trees.

Butternut squash soup
Bream wrapped with veggies in puff pastry
Potatoes, veggies
Sticky toffee desert

On our bed was a little note, attached to a mahogany leaf:

These leaves are from the *Trichilia emetica*,
better known as the Natal Mahogany tree.
Traditionally the leaves are said to have special
healing properties that may aid insomnia.
The staff and management at the *Islands of Siankaba*
wish you a most peaceful and restful sleep.
Moone Kabotu
Good Night and Sleep Well

It was incredibly romantic here!

During the night we did hear a hippo yawn, gurgle, and splash right in front of our tent. And listened to crickets and river water all night long.

We awoke to an amazing dawn chorus, and immediately started birding from bed in our elegant tree house. There were pairs of bright black and white tropical boubous chasing each other through the branches of the trees out in front, with their loud ringing crystal calls and answering "aack"s. The Hueglin's robins were down below the house in the leaves, scrabbling around and singing their robin thrush songs.

Hot coffee and tea were delivered at 6:30am, and decadence reigned. We took our time emerging from bed - it was really quite cold! Taking in breakfast at the main lodge, out on the decking by the river's edge, we birded as we ate. At this point we were up to 44 species of birds (eight lifers), when suddenly Jim pointed out a very bizarre looking bird making its way along the shallow water near the riverbank. It was a duck/heron/darter-like thingie, with an unusually thick bright orange bill. Jim excitedly thumbed through the bird book, to discover that we were looking at the rather elusive and rare African finfoot. Thrilling!

We spent a little time wandering about our island, exploring the nature trail through the deep trees, finding birds and keeping our eyes peeled for crocodiles as we'd earlier seen several from the boat, basking on the sandy river banks of our island. Soon it was time to meet our guide, Doctor, on the mainland, and we motored over in the boat launch, again spotting a crocodile en route. Glad we hadn't actually seen one when we were on foot!

Doctor was given his nickname as a child when he would play with the stethoscope of a medical doctor at the mine where his father worked. He lived in the small village after which the safari camp is named, Siankaba, and offered to take lodge guests for a village tour, a popular activity for people staying here. It was simply fascinating to see, in detail, such a different culture from our own and, we couldn't help but think, healthier in many ways. We loved spending time with Doctor; he was a sweet endearing soul, filled with knowledge which he clearly loved to pass on, and a gregarious chatter box who would demonstrate a point by drawing in the sand with his walking stick.

When we arrived in Siankaba, the young children ages three to six from the small local school came outside to say "hello" and pose for pictures. Doctor pointed out later which one was his son, a very sweet little boy. All of them were sweet, though, and it was really special for me to be surrounded by so many adorable little kids. The teacher eventually rounded them back into class, and we could hear them singing their English lessons, reminding us of the movie, "The King and I."

Doctor showed us how the huts were built from mopane sticks and bark, and then covered with termite mound soil. He drew pictures to illustrate how young children lived in huts with their parents but after the age of six moved to separate dorm huts for the boys and girls, explained about the bride price (gosh, you had to pay it back if you got divorced!), discussed the local politics and structure of the chiefdoms (five of the 14 local headmen were female), and even demonstrated how the kraal - stick enclosures for cattle - were moved across the fields to allow for even fertilization. This latter concept is something that American organic farmers are just now beginning to realize is a brilliant idea.

Being the dry season, no crops needed tending and Doctor mentioned that this was a time of relaxing for the villagers compared to the very busy wet season. Therefore, it was not surprising to see a group of young men hanging out and drinking beer in the middle of the day. They were obviously having a great time, and offered a swig to me. I did take a drink but I must admit to not really enjoying it. The container was an old dirty plastic jug, and the "beer" was sort of thick brownish gray and lumpy. Uck. We were shown that this town of about 12 families had one car, one TV, and one radio, the latter both run by a car battery charged up at the lodge for a small fee. The older children, ages seven to 16, walked to school in the next village over, called Mapena. This was the central town in the catchment area of 14 villages and had the one upper school for about 400 kids and one health clinic. Both of these were staffed by the government; the teachers and nurses were assigned to a particular village and housed next to the school. Again, it was great fun to spend time with the kids, all of whom clamored to have their photos taken. We appreciated one of the teachers showing us what the 7th graders are studying - pretty much the same stuff 7th graders study in the US. It was "health week" and all kids were being de-wormed and treated for Bilharzia, which is caused by drinking the river water from slow pools. There was also an athletic competition going on, choosing the team for the regional championships, and we were entertained to see them practicing the high jump: Jim's sport in high school.

Doctor walked us to the edge of the river, and we met another camp guide with a mokorro, a small wooden canoe. We climbed in, sat on the low chairs they kindly provided, and punted down the river. It was a little hair-raising when we barreled over the class four rapids (just kidding, they were just tiny ripples), but a pleasant way to experience the Zambezi River. We pulled up at the boat dock and were told that there was a picnic lunch awaiting us on the island. All we had to do was walk along the nature trail, stay to the left on several forks, and miraculously, we'd see the food to the right.

Well, we did as we were told, walking the length of the island through thick forest and ducking under overhanging branches and although we carefully searched, we didn't see any food on the right hand side. When we reached the very tip of the island, we realized we simply had to go back and try again, making lots of jokes about how panicked we were after such a long walk (1/4 mile), that we should arrive back in camp tattered and dirty, covered with sand and leaves, and crawling with hunger. Or better yet, simply send our bedraggled and bitten hats down the river, to slowly float past the main lodge. We really laughed at that. Guess you had to be there. But anyway, we came across Doctor, who had come to warn us that through a small kitchen snafu the food hadn't been delivered quite yet and that he would assist us with some bird watching along the trail.

When lunch arrived, it was quite the feast by the riverside, on a table with picnic benches nearby, reclining chairs and a hammock. The food was on a small square serving tray with four compartments: rice with cilantro, corn, peas; baked tomatoes on a croquette with feta; breaded bream with mustard sauce; green salad. Bottled water, cokes, and bread baked on a stick rounded out our yummy and romantic lunch.

When done we were told to simply leave all the picnic fixings, and so we slowly wandered back to our room, where Jim rested and I got a magnificent massage on a table out on our deck overlooking the river. What a life! I was feeling particularly spoiled. We finished the day with a sunset cruise along the river, drinking Amarula and watching many gorgeous birds including an African black rail. Three female kudu stared back at us from the river's edge.

Dinner was again a sumptuous affair:

Lentil soup
Beef (spicy sweet potato for me)
Butternut squash
Beans in a delicious sauce
Iced coffee in tall port glasses/warm chocolate soufflé

Later, back in the room I called Mark since there was cell reception, and was overjoyed to hear that he had gotten his passport finally, and was going to be able to join Jessica in Honduras for a shortened trip. It made my day to know all was going well with the kids back home.

On the fluffy bed, inside the mosquito netting, was another lovely note:

> Our signatory bird, the trumpeter hornbill, has a very distinct call. You may have heard a cry almost like that of a human baby during your stay. This in fact is the song and call of our resident friend.
> Below is an African proverb that we have given to you as a form of inspiration and future guidance from our local people. It is said that a proverb holds great wisdom that has been passed down from elders in a tribe for many years.
> "Learn from the hornbill, the bird of unconquerable hope.
> No matter how bad the drought, no matter how desperate the famine, the hornbill always holds its head high, its beak pointed towards a better tomorrow.
> Never be like a crow whose ugly beak points earthwards in pessimism – be like a hornbill, my child."
> We trust that you will take these wise words with you when you leave and remember the Islands of Siankaba with great affection.
> Good night and sleep well.

It was a significant note, the words upon which we took with us into delightful sleep.

After another pleasurable night of river sounds, the next morning we were up with coffee and tea delivery at 6:30am, then took the launch back to the mainland for our checkout. While doing the paperwork, Lausanne, one of the hostesses, asked if we'd like to take some bottled water with us for the day's activities. I mentioned that I had been simply drinking water from the tap in the room and she was horrified. "But that's not filtered!" she exclaimed.

"Well, that hadn't really been carefully explained," I mumbled, all the while feeling a little sheepish. I mean, the first thing you learn about traveling to developing nations is NOT to drink the water. I quickly changed the subject, and it was then time to say goodbye to the Islands of Siankaba, which we had truly enjoyed.

And now we were off to experience Victoria Falls again. Conrad, a guide at the Islands, was our host and most helpful. We passed the small Livingston Park along the way, where we saw two giraffe and two impala leaping. Conrad told us that the black rhino re-introduction, put in place after the rhinos had become extinct in Zambia, wasn't going too well. They had purchased four rhinos for this park a few years back, and we had seen them from the helicopter the last time we were here. One of them was pregnant and so, after giving birth there were five rhinos. Over the next three years, however, one of them drowned, one got an illness and died, one died in a fight, and this past month the remaining two were shot by poachers. We were happy to hear that the one rhino who survived the shooting was the baby who had been born in the park. Regardless, there is a lot of on-going discussion about the best way to move forward. The most likely opinions are that this park is too close to human habitations and less safe from poaching than the more remote parks. Next time they are planning to re-introduce them into the LZNP and in fact we met some of the rhino researchers when we were there later in the week.

A most definite highlight, we took exhilarating microlight flights over the stupendous Victoria Falls, bundled in warm padded flight suits. This was really the best way to experience the falls, especially in a year with so much water as the mist can obscure the view from the ground. We were in separate planes but could see each other at a distance. It was a fantastically fun 15 minutes. I loved flying directly over and really feeling the power of the mile-long falls, with implausible amounts of mist rising up, and seeing our first elephants of the trip, from the air.

Conrad then drove us over to the falls so that we could walk the pathways to view the falls from all the overlooks. We were provided large plastic ponchos to wear, and boy did we need them! There was so much mist and wind that at times it appeared to be raining UPWARDS. Once again I was very moved by the sight of those falls, and the vertical rainbows. I felt Kalpana's presence again, shedding some sweet tears for missing my dear friend.

From the serenity of nature to the chaos inherent in traveling, we made a mad dash to the airport as we hadn't really saved enough time to do all that needed to be done. Lausanne had been very helpful in calling around to the airports to see if our baggage had arrived. She was told the night before, that our bag was to arrive at Livingston airport in the morning. We were thrilled hoping against hope that it was Jim's bag and not the box with mouthwash, and so were anxious to get to the airport. First we needed to stop in town at Barclay's Bank, literally the only place we could get some US dollars. We only got $300, as they didn't have very much US money in the bank, so we couldn't get more. However, we figured this, along with the extra money we had brought for shopping, would cover the tips and our exit visas. We then rushed to the airport, ran to the BA office, only to be told that Jim's bag hadn't arrived. The woman in the office had no idea what we were talking about. "Perhaps it got put in some other office, or it could still be in customs," she suggested.

Aside from the frustration of the series of miscommunications, Jim really wanted to get some clothing now that his bag was not actually at the airport, and wished that he had known he was going to have to buy some when we were in town earlier. But Conrad insisted that the 25 minutes remaining until 12:20pm (the time that Air Zambia said we HAD to be ready to go) was sufficient for Jim's shopping. So they zoomed back to town, while I ran from room to room, knocking on doors and asking if they could possibly have my luggage. I finally made it to the customs area where I kept getting re-directed to different offices until I came to a room with a curtain instead of a door. Not being able to knock on a curtain, I simply pulled it aside and poked my head in, simultaneously saying, "Excuse me?" And what a scene met my eyes! Two huge scary looking black guys, stopped in mid-sentence, appearing to have been interrupted from some really nefarious dealings, were casting seriously murderous looks my way. I felt really small and vulnerable. "Do you SEE any bags in here?" one of them barked, and with a tiny, "Well, no …" I quickly slinked away. Yikes.

Jim and Conrad ran up to the departing area at 12:17pm with three minutes to go. "What's the rush?" I joked, but luckily we made our flight. Jim recounted the ridiculously funny time shopping with Conrad who thoughtfully kept trying to get a better rate of exchange for a pair of UNDERPANTS, while Jim kept explaining that there really wasn't time. Jim didn't care about the amount of kwatcha, the local currency, that he was given back for each $20 he spent although Conrad clearly cared, as he was the one pocketing the change! Jim managed to find two pairs of undergarments, one

T-shirt, two short-sleeved safari shirts, and two video tapes. Not bad for a spot of African power shopping!

And we were off!

As we landed at Lusaka International Airport, I must admit I shed a few tears of joy. Dave Cumings, owner of Chiawa, met us yet again and helped us negotiate through the airport. He was so helpful, even driving us over to the Proflight hanger for our small plane, so we saved the usual $8 exit visas we would have had to fork over if we'd walked on our own. We really appreciated that, given our current money troubles. He also had a bag of four video tapes that Jenny had graciously purchased for us. They are really such nice, nice people.

At the plane, we met up with Jason, the owner our other favorite camp, Sausage Tree (and co-owner, with the Cumings, of Old Mondoro). We noticed another fellow standing by the plane only to exclaim with fun and joy as it was Moses, a guide at Sausage Tree with whom we've spent fascinating game drives during two previous visits. Moses is always so kind as to remember us, even recounting some of the exact experiences we've shared. He was flying back to the bush after a short holiday with his family in town. So it was a great group that flew into Jeki, while I sobbed literally the entire flight. The return to the escarpment, the braided Zambezi River, the green grasses and towering trees, elephants in the river and guinea fowl on the runway was emotionally overwhelming to me.

Our jeep was waiting for us and Moses drove us to camp. We saw 12 species of birds, all of which were beloved from previous years … the gray lourie plaintively calling "go away!," the majestically soaring fish eagle, and the brilliant black fork-tailed drongo. Ah yes, my symbol of Kalpana, welcoming us to the LZNP.

I thrilled to see our first elephants, a troop of baboons with babies riding on mothers' backs or being held to mothers' tummies, and a large male making an almighty sneeze after a frightening grimace. The baobabs moved me to tears again as did the palm trees, massive spreading winter thorn trees, mopane trees, and sausage trees, the latter mysteriously – to me – devoid of hanging sausages.

Chirapira Mountain peeked above the escarpment, and Jim won the first tsetse bite.

Per the usual routine, on arrival at Sausage Tree Camp we were handed warm face cloths, and met by our hosts Dave and Tania.

We drank our Amarula sundowners on the deck, with yummy snacks and an incredibly radiant orange sunset over the escarpment. I snuck a visit to the guest loo, and giggled when I saw our coveted wooden guinea fowl family.

And then we heard hippo calls! Many of them in a row, loud, right out in front of the deck over the river, so enchanting. Dave explained that the local resident hippo we'd remembered from our first year, Frank, was still there, along with his side-kick whom we met two years ago, Stein. But there were several more hippos hanging about with Frank now, and the names were variously given as Furter ("Frank Furter"), Sinatra, and Incense and Myrrh ("Frank, Incense, and Myrrh"). Boy, the sense of humor in the bush is an intriguing thing to behold.

After drinks it was time for our first night drive of the trip, and it was a fantastic one, just the two of us with Moses and a spotter. With the Southern Cross, Jupiter, and Venus overhead, we stopped to see our first genet.

We then examined a nightjar who stayed still enough for us to determine that he had patches of reddish color to the sides of his bill, hence the rufous-cheeked nightjar.

There were a few fireflies flitting about, and then we saw: genet, genet, civet. And then … KAINGO!

This female leopard was quietly walking along through the low grasses right next to the jeep, giving us a perfect view of her gorgeous coat. What a fantastically beautiful animal. She paused a bit and turned back, then disappeared behind a tree. Suddenly she burst out from the other side of the tree, dragging an enormous male impala by the neck. Apparently she'd killed him earlier and was deciding where to stash him. She munched on him a bit, poking her head into the rather gory bloody mess of his abdomen. The eyes of the impala were staring right at us, with legs akimbo, as the leopard had her head inside, grabbing at entrails. Wow, "nature red in tooth and claw," for sure!

She tried pulling the carcass up into a nearby tree, but fell to the ground. She then spent some time clearly thinking it over, glancing up at the tree, then down at the carcass, and back up at the tree again, sizing up the situation. She gave up after a while and wandered off, flopping down to lick her paws and relax.

An extraordinary experience!

We eventually drove off to leave her in peace, and saw a genet, followed by another genet. This reminded me of some previous trips when we'd see so many genets at night that we'd just call out "four!" or "six!" or "nine!" depending on the count. We also saw a giant eagle owl in flight, perching and dipping his head while hunting from the branch.

Back home to warm towels, a chat around the fire with the other guests, and then dinner at the long table. Honore, the chef, delightfully described the menu for us.

Broccoli soup
Beef (or quiche for me)
Sweet potatoes
Potato wedges
Lemon meringue pie

We liked speaking with two youthful guys who were traveling with their folks, down from their property in the north of Zambia. They were in university in America, but had grown up in Zambia. They say that their property has the highest number of species of birds in Zambia … perhaps we'll have to go stay in their lodge some day.

Before long we headed off to our chalet, and once again, what an incredibly romantic, elegant chalet it was. The round room had smooth mustard colored concrete floors, thatched walls, and a tall white canvas tent overhead. The luxurious tall bed was opulently draped by mosquito netting, while the kerosene lanterns gave a warm glow over all. Lotions and potions were provided in the loo, and sweet little art arrangements and local flowers added a decorative touch.

We fell asleep to the sound of hippos at long last, together with crickets and the soft tinking of tiny frogs. During the night the hippos were often loud, burbling and guffawing to my heart's delight, and a large animal (an elie?) munched right next to our tent walls.

At 6:00am, coffee and tea were delivered to our room by Simon, who kindly remembered us from our last visit. After a yummy breakfast on the deck over the river, we were off for our safari drive, and again it was just us with Moses. We had a pleasurable day luxuriously spending inordinate amounts of time focusing on birds and figuring out their identification from subtle details. We had to be sure exactly which woodpeckers we were seeing, and whether the orange-breasted bird flitting about was indeed an orange-breasted bush shrike, or in fact the very similar gray-headed bush shrike.

Fish eagles gave haunting cries overhead, woodpeckers rapping, red-eyed and Cape turtle doves calling, "Work harder! Work harder!"

We learned about some of the trees, a number of which look different at this time of year. For example, the rain trees had pods all over, whereas previously we've never seen them with pods as the August winds blow them all off. Jim also pointed out a low bush with bright yellow, interestingly shaped flowers, and suggested that it was the famous scrambled egg tree. I vehemently disagreed that that could be the case, and as always when I vehemently disagree with Jim I was soon proved wrong, as Moses pointed out the "scrambled egg bushes." Harrumph. I was teased mercilessly about that for the rest of the trip, as Jim would say, "Now, what are those YELLOW bushes called?"

The dambos were filled with deeper water at this time of year, and the water hyacinth had many more purple flowers. In fact, everything was more lush and green. This was pretty, especially watching the large shaggy waterbuck actually splashing through the water, but the drawback was that it was more difficult to see the animals as they are more spread out with so many sources of water all around.

We watched stilts and wagtails in the marshy dambos, and saw our favorite little black crakes with their improbable bright red legs and orange bills. We watched the different styles of the sacred ibis and the great egret fishing, the former constantly pecking its decurved bill down into the water, the latter standing completely still until spotting prey and then stabbing once with the long sharp beak. There were several squat herons that could either be the more common squacco, or the more rare Madagascar. We wouldn't leave until we were sure of their ID. They were all squacco, unfortunately, as we had wanted to see the rare one we'd not yet seen on the trip.

We gazed at a brown-headed kingfisher on a branch right over our heads, while listening to the puff-back shrike yelping, Hueglin's robin singing, and the hadeda ibis cackling. A hippo head bubbled up out of the dambo water right beside the jeep, stared at us, then bubbled back down under the dark water. Wow!

We had now seen 90 species of birds on the trip (ten lifers), and enjoyed the marvelously vivid white, black, yellow, and red of several graceful saddle-billed storks. At this point we had an experience so tragic that it's difficult for me to describe (don't worry, I'm partially joking about it being a tragedy). We were sitting near some trees, examining another orange-breasted bush shrike, when Moses turned his binoculars to the tree next to the one we were looking at. He mentioned that there was another bush shrike in that tree and then exclaimed, "but I see some RED ... "

And I knew EXACTLY what he was talking about! I'd been studying my bird book and knew that there was a dazzlingly stunning bird called the "gorgeous bush shrike" that looked just like the orange-breasted, but with a red throat. Jim called out that he found it with his binos, and started gushing about it being the most stupendously beautiful bird he'd ever seen, but I didn't know where they were looking! I kept asking, frantically looking all around the area they were pointing at when ... you guessed it.

The bird flew away.

I never saw it.

The gorgeous bush shrike. A life bird even for Moses. A bird that no other guides we spoke to had ever seen.

And I missed it.

Aaackckck!

Well, but we all must move on, and I managed once again to see beauty in the world. Jim was pretty kind, actually, not teasing me. I think he worried about a return to the stolen-hot-water-bottle incident.

It was fun to see our first warthogs, about which time we stopped for a coffee and cake break at a scenic overlook on the Zambezi, where hippos were submerged out in the river. Moses found a good "Douglas-free zone" for marking our territory, Douglas being the local large male lion. As the drive continued we watched a duet of two Hueglin's robins singing to each other in a winter thorn glade near a majestic sausage tree. An African hoopoe was posing up in a tree, arrow-marked babblers were babbling, and tree squirrels frantically ran around us.

I was happy to recognize a group of white helmet-shrikes flying by, and to see our first elegant bateleur eagle regally soaring overhead.

In fact, now that we saw the bateleur, Moses suggested that we try to find the leopard's meal from the previous night, as perhaps that's what the bateleur was searching for too. We drove to the tree where we'd seen the aborted attempt at climbing, and Moses showed us various drag marks and footprints in the sand to try to follow the leopard's progress with the impala. We drove again through the field of lavender we'd driven through the night before, and smelled that same delightfully intoxicating scent. We hunted back and forth, under the hot sun, into and out of bushes, examining all the branches of all the possible trees for impala bodies.

Suddenly we realized we were no longer alone, as vultures now began to appear overhead. Eventually there were 17 of them (white-headed, white-backed, and hooded) soaring, circling, diving, and landing on various nearby trees. We knew we were close; the turtle doves called at us to "work harder."

And then, Moses believed he could see the impala. We spent the next half hour in deep discussion, arguing about whether indeed what we could see up in the tree amongst the foliage was a dead impala or not. We each drew diagrams for each other, and stared and stared through binoculars. We were still unsure even after all that work, so Moses suggested we try to look at it from another angle. The brush was too thick for us to drive any closer at this section of the tree, so he backed up and we tried various times to get close from a different approach. It was not easy, as the undergrowth was so thick, but finally it appeared that we were going to get close enough ... and we patiently held our binoculars at the ready ...

And Moses suddenly stopped the vehicle, pointing downwards.

Kaingo!

She was sitting under a tree nearby, delicately licking her paws and gazing skywards at the circling vultures.

It was quite incredible that we hadn't seen her during all this time. We joked that she'd had to move her tail several times, to avoid having been run over by our jeep. Anyway, this solved the mystery as we looked in the tree above her and could clearly see a dead impala. I wonder what that other "thing" had been in the other tree ...?

Anyway, we happily watched her – with her gleaming green eyes and her luxurious stunning coat – for quite some time.

To top it off, on the way home we saw an elie mom with a tiny baby, munching on seeds.

Lunch was a special affair, out IN the river with Jason and Tania. We were brought to the table by boat, and again thought it was quite unique that the table and chairs were all set in the water, so that our feet could dangle in the cool river underneath as we ate. The food was fantastic including a zucchini quiche prepared just for me. We sipped on a pitcher of Pim's and lemonade, a perfect beverage in the moment. It was really a special treat being the only folks in camp on this day, as we got such personal attention.

Back in our room we had wonderful outdoor showers, and then a bit of a tearful "goodbye" to Moses as it was time to move on to Old Mondoro bush camp.

Roelof was in the boat to take us down river for the one hour ride, and we had a happy teary "hello" to him. He is one of our absolutely favorite guides. We had spent hours and hours alone with him on our previous trip and loved the way he showed such respect for all the wildlife, and reveled in each animal we saw. We had missed spending time with him last year and also with his enchanting wife, Helen, so had requested a good long stay at Old Mondoro this year. Earlier in the season when Helen and Roelof were setting up the camp, there had been an unfortunate explosion and fire accident, whereby Roelof was extensively burned. We were shocked when we had heard this earlier in the week so were thrilled to see him back in camp and doing well.

Because his skin is so sensitive while it recovers, however, Roelof needed to stop guiding and so he and Helen had turned in their resignations. Grant and Jason had found a new couple to guide and manage the camp, but they had still not received their work permits. Roelof needed to leave for South Africa for another operation the following week. It was a tricky operation and Helen wanted to be by his side. So, Jenny asked her young friend Katie if she'd run the camp for a while, until the new couple received their work permits. Katie arrived the following afternoon, and so we got to watch her training from Helen, and her interactions with her first guests. It was exciting to be in the inner circle. We were the only guests at Old Mondoro for our first three nights there, which ended up being the last three nights that Helen and Roelof were there; on our fourth night a new family of guests arrived for Katie to work with.

In the meantime, however, I was quite teary-eyed about the terrible accident, and how awful it must have been for Helen sitting by the bed of her beloved, not knowing if he would survive. When we pulled up at the dock at Old Mondoro, I grabbed Helen in a bear hug, hardly able to speak for the tears. We got settled in, Helen being her usual sweet solicitous self about the tragedy of our lost baggage (imagine being so concerned for us, after what she's been through). The main gathering area of the camp is right on the river bank, a pretty slab of smooth green concrete with lofty beams holding up the canvas roof. Very simple comfortable furniture for relaxing, and otherwise open to the wilderness. We had Mosi beers and snacks - coconut cake and mushroom pizza - but I didn't feel like eating. Hmmm, that's funny, that's never happened before ...

Helen took us to our perfect chalet, right on the point of the intersection of the Zambezi River and a little side channel, so our room looked out on both! The side channel was filled with water hyacinth and water birds, and the Zambezi was filled with hippos. Heaven. The chalet is very simple, as this camp is so far out in the bush: thatched grass walls and canvas tenting, atop a smooth concrete floor. There is also a little balcony pointing out towards the channel with chairs for relaxing, a loo surrounded by grass thatching, and a shower with a tree trunk forming one side. When you desired a hot shower, water was heated over a fire, brought to the chalet by wheelbarrow, and hoisted up into a bucket in the tree.

We unpacked such items as we had to unpack and I used the loo, noticing my tummy was aching a bit. Hmmm. Thinking back to my drinking the water at the Islands of Siankaba, recalling their comment about it's not having been filtered and the typical two-day incubation period for intestinal bacteria ...

Nonetheless, we went out for an awesome game drive, with Steve as guide, Levy as spotter, and Roelof as assistant in the back of the jeep. Both Chiawa and Sausage Tree guides were taking turns coming down to Old Mondoro in Roelof's convalescence.

The first wildlife we came across was "Limpy," a titanic elephant bull so named because his back leg had a broken knee, and so his hindquarters had not fully developed and he walked with a pronounced limp. But nevertheless he was a magnificent animal, in the prime of his long life, and we were simply captivated by watching him. He reached way up into the acacia trees, pulling down branches to shove into his mouth, even consuming the thorns (which are edible by virtue of being slim and hollow). Steve explained how important the elephant feeding process actually was to a complete ecosystem, as the impala nearby fed from a branch he'd dropped. It is an important source of browsing food for the impala during the dry season, as the browse line on all the trees is now at the limit of their reach.

It was fun to have all three photographers on board - Jim, Steve, and Roelof - snapping away like mad, while listening to the jabbering of Meve's starlings, turtle doves calling, and hippos garumphing and laughing. At one point Limpy appeared to be coming directly towards us, but it was only to place his pendulous head over a low slung branch, to scratch his neck by rubbing back and forth across it.

And then it was on to the kudu kill!

Steve told us about what we'd missed earlier in the day, pointing out where it all had occurred as we drove. He and Levy had dropped the previous guests off at Jeki Airstrip, and on the way back to camp they came across an unbelievable sight: two already rather full male lions pulling down a full-grown kudu bull. It was thrilling for them to watch, but otherwise confusing: normally two lions would not be sufficient to kill a full-grown eight-year old kudu, which are really massive antelope. But they succeeded in killing him, which was quite sad as he'd been a gloriously regal animal, with immense corkscrew horns. Roelof wondered if he'd had something wrong with his ankles, as he's seen in other kudu in the park over the past years. He was trying to investigate the situation, as no one really knows what's going wrong with the kudu ankles and perhaps the weak ankles could explain how the lions had the advantage over the kudu.

In any case, after describing the scene, Steve pulled up to the kudu carcass and we were amazed at its size. The spiraling horns with three full turns were exceptionally large and handsome. My first comment was, "I want those horns!" However, Levy had already spoken for them. Not much of the antelope had been eaten; its stomach was lying on the ground next to its open rib cage, and some of the meat was missing, but his head was still quite intact. No vultures had gotten to his eyes yet. He was a really beautiful guy: kudu are colored a delicate gray with a surprisingly pretty white stripe across the face under the eyes. I felt sorry for his death.

Steve drove the short distance over to the dry riverbed nearby to see if we could spot the murderous lions, before taking the chance of hopping out of the vehicle to inspect the carcass. And this turned out to be a good plan: there were two lions lying in the sand quite close by. However, they didn't look too formidable, as they lay there in the sand with bloated bellies, snoring and urinating without bothering to move. What lazy bums! We learned that these were two of the three lions who had swum over from Zimbabwe earlier in the season, and had set up a "coalition." There was also a lioness involved, but she was the property of the dominant male and so the other guys were forced to stay away from her. The two we were looking at were the two sub-dominants; perhaps the big guy was hanging out somewhere with the lady. Steve explained that one of these two, along with the big black maned male, were the ones who brought down the kudu. Looking at them now, with their rather scruffy manes and scarred pockmarked faces, it was hard to imagine. Steve said that after the kudu died, they had actually licked at each other, sort of like celebrating a triumph, rather than diving right away into the kill as normally happens. It's pretty cool that they have developed this coalition and cooperative behavior.

We drove back over to the kill, and everyone, except me, got out to inspect the carcass and take photos. Again, it was fun to be in the inner circle, just us with a bunch of excited guides. Roelof examined the feet and ankles and didn't see any problems. They examined the whole animal and couldn't figure out why it had lost the battle with the lions.

Then it was time for the guys to take photos posing with the kudu. The photo of Jim proudly holding the head and horns looks like HE actually killed the animal. When Roelof was using his camera with a very long lens, he kept backing up and backing up to get a good shot. He was nervously looking over his shoulder, and we were all teasing him about whether it would be smarter to simply put on a different lens, rather than getting attacked by a lion protecting his kill!

As we'd spent so much time with the kudu and the lions, we missed actual sundown and had to have "sun-goners" instead. We needed to be careful to stay near the jeep as lions were about. So I peed next to the back wheel of the jeep, and felt that I really needed a real loo for my aching tummy. Hmmm. I nonetheless allowed myself a touch of Amarula, and we continued on our night drive.

Nightjar. Giant eagle owl. Genet.

And then …

Kaingo!

She was very skittish, hiding behind a termite mound, peeking out from one side but then moving to the other side when we shined the light on her … and when we shined it on the other side, moving back again. We didn't want to bother her too much, so after following for a while, and making sure to get full face-on photos, we left her alone. Later, Roelof determined from the photo that she was a new girl in the neighborhood, hence explaining her nervousness around jeeps. They've named her "Nsoni," which means "shy" in one of local languages.

I was feeling pretty sick by now, and asked that we start to head home but first we stopped to check on the kudu corpse and discovered that it had been moved and further eaten. We checked the lions, and this time the dark maned, larger male had joined the two others, with a really bloody face. Guess we know who was eating at the carcass recently!

We saw one photogenic white-tailed mongoose, literally posing for us, and continued onto camp. On arrival I simply had to run to the chalet for the loo. I won't go into the gory details of my dysentery but let's suffice it to say that I was ill all night with a high fever and many trips to the loo. I remembered that the most important thing was to stay hydrated so we had Helen mix up re-hydrating solution for me and I proceeded to drink four liters of it over the next 24 hours. I also, thank goodness, had a prescription for ciprofloxacin (the anti-diarrhea broad spectrum antibiotic that we'd been bringing with us each year) and started taking it right away. Most of the night

I slept with both hot water bottles cuddled next to me, but at one point I remember stumbling to the loo and being so hot that I ended up lying down on the cool concrete floor to cool off. Jim found me like that and was a bit worried as the floor was not cool, it was freezing. So he put a blanket and pillow under me and was very careful caring for me. He even had to wash down the bamboo walls of the loo during the night and yet he still loves me.

At one point, when Helen was checking on me and bringing more hydrating solution, I was embarrassed about all the trouble I was causing her. I started to say "I'm sorry," but she pointed at me and sternly stated, "If you are going to say you are sorry, I am NOT going to accept it." What a woman.

Morning came and I was slightly better, and by the time it was early afternoon I was actually able to get out of bed. Jim had gone over to the main room for lunch and told me that Grant and Lynsey had arrived, with their little 18 month old boy Scott whom I had not met yet. They had brought Katie to Old Mondoro to learn the ropes from Helen. I slowly got up and dressed, and carefully walked over to the main tent to say a tearful hello and to hold Scott. I felt quite weak but seeing all these wonderful people cheered me up.

I returned to our tent for another nap and we agreed to meet at the main tent at 4pm for afternoon tea. I had told Helen I thought I should only eat rice so the staff made me a sweet little tower of rice, decorated with a sprig of parsley! I cried when I saw it, and took photos. So tremendously sweet!

I felt good enough to sit in the jeep, so we went out for our afternoon game drive. We'd missed watching the lions fighting over the kill in the morning, and now the kudu was literally just a bizarre skeleton of ribs, a spine, and a skull, and the lions (all four together now) were insanely bloated and lying in the riverbed, appearing completely unable to move. There were also vultures all around, mostly hooded, standing on the sand, hanging around the lions with the intention of eating their feces.

We enjoyed the sight of a large male kudu, alive, and several side-striped jackals in their monogamous pairs relaxing in the grass.

Now that my dysentery was under control, the tsetse bites on my ankles were really beginning to bother me, along with my canker sores. There were no sausage fruits on the trees, however, which I'd noticed earlier, as it was too early in the season. So much for my holistic treatment for tsetse allergies! And furthermore, since I'm complaining, I may as well add a comment about the rumbling of my stomach. It was as loud as a freight train. I felt as if I were communicating to all the animals in the bush, I just didn't know what I was saying to them.

We drove the jeep across the sandy river bed, inhaling the delightful sage aroma as we passed the lush green bushes on the riverbank. We had our sundowners up on the Jeki plain, with stars coming out – Southern Cross,

Jupiter, and Venus. I restricted myself to mostly water with just a sneak of Amarula from Jim's plastic water bottle cup.

And then the night drive began.

The funny thing is that we only saw two genets, a few fireflies, and a hippo decorated with several clumps of hyacinth on his back, yet it was still lovely and intoxicating to be out under the stars, driving through the African bush, seeing what we could see. At times we simply sat quiet in the jeep, down by the river, listening to the tinking frogs and the crickets, with firefly light reflecting on the hood of the jeep.

Reveling in the African night.

As we drove up to the camp, decorated all over with brightly lit kerosene lanterns, I fell in love with Old Mondoro all over again. We were handed glasses of sherry and warm towels, and invited to the fire circle by the river for snacks and more drinks. Of course I cried. I didn't drink the sherry, still being worried about my (loud rumbling) tummy, but did sit by the fire and chat. I joked about having loved my water bottles so much while I was sick that I had named them "tweedle dum" and "tweedle dee" while Jim pointed out that a much more appropriate name, based on my obvious desire for them, would be something more like "Fabio" or "Robert Redford."

Dinner was out on the riverbank, on a table set with white linen, crystal, and china, and with bright candles set into a metallic crocodile candelabra. The napkins were carefully folded into the shape of a man (with a spoon for the head!), and there was a warming brazier at our feet under the table.

Potato leek soup
Garlic bream with veggies (just steamed bream for me)
Rice pilaf (just steamed rice for me)
Chocolate mousse

The night was magnificent ... I only hoarded one hot water bottle, didn't have to get up for the loo, and was able to appreciate the loud hippo slosh-eating right outside the tent, along with the loud raucous hippo calls.

And we heard lions calling at dawn.

The next morning we were awoken at 6am, to the sounds of the hot water being added to the bucket in our tree so we could wash our faces with warm water. We met at the main room for coffee, tea, and toast over the fire by the river, and were off for an early game drive. Since we'd missed yesterday, I was anxious to get going. It was quite chilly, but blankets were provided to cover our legs and I had my jacket/hat/gloves while Jim had his fleece from Wolwedans. The delicate morning light filtered through the trees, and we watched extensive herds of impala leaping and cavorting about.

We then saw a female kudu with a tiny baby, and were able to watch them quietly for some time. The adult was very cautious before crossing the sandy road, looking tentatively in both directions, but when the baby came to the same spot he just blindly skipped across, following mom and oblivious to the rest of the world. It was fabulous to watch the baby nurse right next to us, mom and baby so relaxed. He banged really hard on his mom to get at the milk; she eventually became too irritated and walked off.

We finally saw our first lilac-breasted roller of the trip, and saw many more of them as the days went by, along with a three-legged warthog (must have lost his leg from a snare, poor guy). We had long discussions about the tragic results of snares, and my solution is that we've simply got too many people on this planet so we need to start culling some.

Driving along the Jeki plain, we saw literally hundreds of warthogs over the course of the morning. The big males have bulky protrusions under their eyes – to protect them while fighting or sparing. We also saw a family of dazzling zebra, standing in the tall grass and staring out at us. They were such pretty creatures, with their brilliant black and white stripes!

We heard some frantic baboon calls, which we recognized as alarm shouts from our experiences last year. We tried to find them in the bush off of the main road, but they eventually quieted down before we caught up with them.

After a break for tea and cookies we enjoyed seeing some watery dambos filled with water hyacinth and exquisite purple flowers.

And then, we had our first real elephant encounter of the trip. There were five bull elies collected around a few acacia trees, and as we drove up the largest one leaned his head against one of the tree trunks and shook the tree while pods rained down all around. All the elies began eating the pods at once, but apparently one of the younger ones wasn't paying attention to proper protocol and was not only shoved by the largest bull, but loudly trumpeted at as well. Guess it was time for some discipline. The younger elephants were expected to stand at the sidelines and wait for the more dominant ones to get their fill first. And after the trumpeting, in fact we did notice that the smallest elephants stayed out at the periphery of the group. Several more elephants arrived to join the first group, and we got to watch more interesting behaviors. One elephant was resting his trunk on his tusk (the trunks weigh hundreds of pounds, after all). Another used its tusks to break off termite tunnels along the trunk of a tree, and then collected the broken-off soil in his trunk to transfer to his mouth for eating. The largest bull made the "close range contact call" – a low rumbling – and they all moved off in formation with the youngest ones behind. One elephant rubbed his butt against a tree, another rubbed his neck. My stomach was rumbling loudly again, and I wondered what it was saying to these awesome creatures.

After the elephants lumbered away, we saw two laughing doves fighting and tussling, and then watched a hefty baboon leap from the tree overhead to a nearby tree for descending, just as Steve had predicted he would. He didn't predict that the lower branch would break, though, and as the baboon made an inelegant flop to the ground we had to laugh.

We heard a fish eagle cry overhead, watched skittish male waterbucks leaping away from us, and listened while a beautiful roller flew overhead and then growled at us. Such lovely birds, but with such unlovely voices.

We then came across three more elephant bulls, this group busily demolishing an acacia. I loved the sound of the cracking branches and the stomach rumbles. One elie crossed his back leg to rest, and we could see the bottom of his foot. It was interesting to see the criss-crossing of wrinkles.

We had to bushwhack our way through the wilderness, as some of the tracks were covered with water at this time of the year. It was fun to do, although a bit uncomfortable from the jolts and bounces, but when we finally arrived at "Long Pools" we were certainly glad to have gone to all the trouble to get there. This was indeed a very long pool, covered with brilliant green and purple hyacinth, bright red duckweed, and contrasted stunningly against the perfect clear blue sky overhead. Five hippos that had been sleeping in the sand at the other end of the pool clambered up and waddled down to the water's edge, continuing their shuffling on into the water until submerged, gliding through the hyacinth as the oxpeckers and egrets on their backs flitted off at the last minute, abandoning ship. Ten tiny little eyes still peeked out at us from under the hyacinth.

A great egret hunted along the shore; fish eagles soared overhead; baboons came down to the water's edge to drink.

And then it was back to camp for a delicious late lunch (it was great being the only ones in camp, as we could come home whenever we wanted), and I was brave enough to try the stir-fried veggies along with the rice. I was finally getting hungry again. Helen graciously carved the freshly-baked loaf of bread (poppy seed this day), as she has done every day since the camp opened.

We went to our chalet for a brief siesta and luxurious shower from our bucket of hot water suspended in our tree. It is amazing how simple pleasures can bring so much joy, if only we take the time to notice.

At tea time, we were devouring delectable fresh carrot cake when we heard the sound of mooing. We had heard the same sound from our chalet a few moments before, too, and asked Helen what it could possibly be. We would have been surprised if the brave and masculine Cape buffalo could sound like a simple domestic cow. However, Helen knew immediately that this was no ordinary buffalo call, but one in great distress. She called Roelof to come over immediately, and requested that the boat driver Andrew rush us over to the island out in the river in front of camp, from whence the mooing was emanating. By the time we actually got going, however, even though we'd rushed, we heard the last "moo" ... followed by a little gurgle at the end. Roelof's guess was that a crocodile had taken him, and the gurgle was his last breath before he was submerged. Since the mooing had stopped, we never did find the buffalo, but it was exciting to travel over to the island to search through the waving reeds and grasses.

Back on shore, we got in the jeep and explained to Steve that we would love to find a breeding herd of elephants. We headed for the albedo woodlands, where the elephant families would be expected to be moving back to the escarpment after drinking from the river in the middle of the day.

And he was right!

We first passed several bull elephants, some with impressive "documents," but didn't stop for long as we were on a mission. We did hesitate at the sight of a tiny bushbuck sleeping under a sickle bush, and thought it was too bad that just because of our arrival she stood up immediately and rushed off from her little sleeping room.

And then, after extensive bushwhacking and jostling and calculating ... we saw ... a breeding herd of elephants!

And an absolutely most magical afternoon unfolded, probably one of my favorite times ever.

15 to 20 elephants marched together in the loosely associated group, snacking as they went. Several small youngsters, and even a tiny baby less than one year old accompanied them, so enchanting.

A young elie scratched himself on a tree; another dug in the mud and then suddenly turned on us and flared his ears in defiance. Such a brave little guy. The littlest one couldn't easily use his trunk, and so instead of stripping off leaves with it, he simply shoved whole branches, still attached to the tree, into his mouth. Another elie pulled up a clump of grass and shook off the dirt before eating.

We heard rumbling and another group of four elephants ran to join the group, trailing a tiny baby trying to keep up.

As we watched and followed the herd's progress, it appeared that they were traveling toward Long Pools. Goodie, we thought, a photo of this extended breeding herd of elephants, illuminated by the slanting afternoon light, drinking at these pools covered in green and purple hyacinth would be a National Geographic kind of shot!

Soon the group arrived at the former bushbuck's sickle bush bedroom, and not only surrounded the bush but began enthusiastically tearing it apart and eating it. Luckily we had already scared the little lady away; otherwise she'd have been in much bigger trouble now. The elephant's ability to eat the leaves of this plant, even with its thick, inedible spiny thorns, is very important for the ecosystem, since when the plant is not kept in check it goes rampant and encroaches upon the environment; no other animals can feed upon it or even penetrate its thick branches. So as we'd learned before, elephant's eating habits are a help to the ecology, not a hindrance.

We must have followed the herd along for over an hour; figuring out where they were meandering to, then driving ahead to get the vehicle in place before they caught up so that we could watch them without the sound of the vehicle disturbing their calm, relaxed behaviors. The matriarch often stepped towards us, flaring her ears as a warning to come no closer. And, as we respected that, there was no further trouble. We listened to the munching, browsing sounds of cracking branches, and were intrigued to see a younger sister tending to a tiny baby so that its mom would have the opportunity to browse.

There was a young bull in the group, about ten to 12 years old, holding back and following at a distance. When a young bull becomes sexually mature, and starts looking at his cousins and sisters in a new light, the females essentially push him out of the herd. He then needs to find a group of bull elephants to join, and although at first he is considered an inferior and has to stand at the outside of any male get-together, the older elephants do teach him the basics of living on his own.

While we watched the elies, we also saw leaping impala and gossiping guinea fowl; listened to Cape turtle doves calling, and saw a slender mongoose running by.

To sit like this, just watching and enjoying a breeding herd of elephants, to simply experience them at home in their universe, was perfect.

I cried.

Eventually we pulled over to Long Pools for sundowners, and our elephant herd hadn't quite made it there. So much for National Geographic. But we loved watching the late sunset light on the hyacinth, and drinking Amarula, and enjoying a troop of baboons. There were so many fascinating behaviors in the troop ... babies on horseback, babies on tummies ... that I even focused my binoculars on them while taking a loo break!

And then, just when the sun had set and its delicate orange glow was gone, a family of four elephants did finally arrive to drink from the dambo. It was an extraordinary sight, even seen in the darker light of dusk. National Geographic probably won't come after us but we had all the pleasure nonetheless.

Such a beautiful, peaceful sight. Two adults, one youngster, and one tiny baby standing at the edge of the water, quietly filling their trunks and then – sometimes in concert – tilting their heads back, curling their trunks around to their mouths, and drinking their fill.

As quietly as they had arrived, they turned from the dambo as one, and slowly melted back into the woodlands.

With Venus beckoning from the west, we hopped on the jeep, ready for what our night drive would find.

Genet on a termite mound.

Genet on the ground.

We shut off the jeep and listened intently.

"Huh-huh-huh."

Kaingo!

We took off in the direction of the sound, and stopped several times to listen again, to be sure of the direction. We heard the distinct coughing once more, but the other times we mostly heard the quiet of the night ... frogs, crickets, elie rumbles.

Staring up at the stars, listening, I was suddenly reminded of the first time I remember sleeping under the open sky, when I was a girl at my summer camp. Our counselor read to us from "The Little Prince," and I remember wondering what a baobab really looked like.

So now we had a new sound in the African night: Cynthia crying by the light of the stars!

At the second kaingo call, Steve (who was helping out from Sausage Tree and thus not as familiar with the landscape) asked Levy (who has the best knowledge of the local area) "Can we get through here?" gesturing to the thick sage brush on the river bank from whence the leopard coughing had been heard.

"No, it's really impossible. Let's give it a try!"

These guys were so adventuresome. Off we went, rumbling through the sand and sage, searching, smelling sage, feeling the excitement of the hunt. Serious bushwhacking, but no one wanted to give up.

However, the kaingo remained elusive. We sat still again ... and then again ... hearing no more calls.

So, it was back to camp, to the warm glow of kerosene lanterns and Helen's welcoming attention. Warm towels, a glass of sherry, and snacks around the fire. Dinner out under the stars with white tablecloth and warming coals in the brazier under the table to keep our feet toasty. I visited the little guest loo, a sweet grass enclosure with open roof and a completely open side facing the river, radiant with lantern light. I could hear impala barking and rutting, and appreciated how quickly I had recovered from being sick.

Cabbage soup
Beef (stuffed butternut squash for me)
Potatoes, veggies
Peach cobbler

Since we were the only guests, we felt as if we were part of the family, and had a wonderful evening filled with intimate conversation. At one point I asked how difficult it was for the staff to prepare a separate, vegetarian, main course for me for each meal, and Helen was quite gregarious on the topic. It was no problem for someone like me, she said, as I simply requested a vegetarian menu (and I also eat fish) and that's not too difficult to cater to out in the bush. However, it's the people who ask for something really esoteric – like vegans, for example, or someone who can't take gluten – that can demand a lot of extra creativity. They found it irritating when guests who had been the most demanding and adamant about their requirements, and for whom Helen had gone to great lengths to find the special foods and have them flown, boated, and trucked out to the bush, who then decided "well, sometimes it's ok if I cheat," and then dove into the main course of absolutely forbidden foods. Helen explained how difficult that is to handle politely.

Later in the evening I asked what was happening with the kudu horns, and Levy explained that he'd gone to retrieve them already, as he wanted to get them before the hyenas arrived as they would crush the skull. He, Steve, and Roelof then discussed the best way to clean off the remaining skin: either boil it – but it would be hard to find a pot large enough, as the horns spread quite far away from each other right at the base – or bury it in the dirt so bacteria can do the work. As the dirt method would take up to a year, Levy opted for the boiling. And then we began to discuss what should be done with the horns when they were ready for display.

Mount them on the front of the boat? So when guests are picked up, they could wonder if they'd accidentally joined a hunting camp instead of a photo safari camp? Perhaps it would be better on the front of the canoes? But then if the head wasn't really cleaned well, it would be rather odiferous as the canoe glided along the river ...

Great idea: make those difficult vegans ride on that boat!

We probably sound like we were being obnoxious, but at the time we thought we were hilarious.

After dinner Helen gave me a truly appreciated gift, a bottle of Listerine. They were going to South Africa the following day and she explained that it was too large to carry anyway.

We then enjoyed a little encounter with Norman, the resident hippo, who was clambering up the riverbank right next to our dinner table, munching on the grass as he came. Helen showed us the best way to walk around him, so that we could make it back to our chalet and a mouth washing extravaganza. I could feel the canker sores improving immediately and no, this was not the placebo effect but the power of positive thinking and good product!

We had another splendid night ... with loud hippo calls and sloshy munching literally all night. Ecstasy!

In the early morning we heard one of the most evocative sounds in the bush. Simba calling.

Loud, visceral, stirring. The sounds of lion, stirring Paleolithic urges.

Later, we heard our 6am "Good morning" wake up, with hot water delivered to our canvas bucket, and quickly got dressed for coffee and toast over the fire on the riverbank.

This day we requested baboons for our game drive. I felt as if we hadn't seen enough of them, so Levy was very thoughtful about finding them for us so we could sit quietly nearby and watch their antics at length. At one point we watched a large guy "rogering" a female, as Levy put it, and a new expression for me, and appeared to be smiling hugely as he was doing it. There were many tiny babies with outsized translucent reddish ears ... some being held to the tummy as the mom ran on her three remaining legs. Levy explained that until the baby is about one month old, it can't hold on by itself. Many older infants were riding piggyback, some stopping to sit in the sun and stare back at us. The animals down here are slightly more nervous about the jeep than in the more populated parts of the park so we didn't get quite as close, but of course that is part of the appeal of being so far out in the bush: the solitude.

At one point we saw a male baboon trying to get near a baby, while the mother was clutching the baby to her chest, screaming and running away. He ran after her, yelling and barking, but she didn't give in and kept her distance. Apparently she was worried that his plan was to eat the baby, which actually happens. But later we did see a large male tenderly carrying an infant on his chest (guess that one was too large to masticate). Levy also explained that baboons had a very strict hierarchy in the troop: some members were considered "royalty" and as that level of importance was inherited, even the tiny babies of the royalty were above all others. Therefore, you could sometimes see a large adult baboon subserviently grooming a small infant. The royal family would also hold up their tails as they walked, in conceit.

We next saw six ground hornbills, big black birds with chunky red beaks, prancing along the trail, picking their way deliberately like little old ladies managing difficult terrain. They emitted a low, guttural "hoot," that sounded like a deep kettle drum. These birds can live to be 60 years of age. They seemed rather plain while walking, just a dull black color, but when flying they spread their colossal wings and the bright white of their inner wings appeared. Apparently, while they hunt snakes, this white was the part of the wing at which the snake would strike; however, this was not serious for the bird as that part of the wing does not carry blood vessels.

Up on the Jeki plain we encountered a dramatic group of zebra, with a baby whose dark mane reached all the way down his back to his tail.

And speaking of drama, some magic was just about to unfold …

Levy pointed out a stunning male bateleur circling up against the suddenly cloudy sky. The delicate lines demarcating the white and black of his outspread wings were elegant against the gray sky, and as he soared and turned we could see glimpses of the ruddy coloring in the middle of his back.

We then heard a raucous "CAW!" which sounded like a crow to me, and realized the sound had come from a female bateleur who had just arrived.

The birds started to circle each other, cawing intermittently. With each call they would drop their vivid red legs and pull in their wings. I'd never seen such behavior before, and Levy and I were gesticulating and marveling out loud, while - luckily - Jim was getting his big camera ready for a shot.

Suddenly, the female DIVE-BOMBED the male. She really appeared to strike him, and it looked painful from below. She did this twice, and each time we instinctively flinched.

And then, the next time they came together, they LOCKED FEET and circled around each other, losing altitude, coming lower and lower, around and around, dropping almost to the ground before letting go and flying away from each other.

It was phenomenal.

Jim got photos (yeah!) and when reviewing them we were able to see that the female was on top during the whole exchange. We thought perhaps we'd seen mating? Or at least a mating display? I don't know why a male and female would be fighting, and since the sexual dimorphism is so prominent (the male has a thick black band on his wings, the female's is much thinner) there was no question that these weren't two males fighting.

It remained a mystery to us, but was a very exciting moment. When they broke apart, they were so close to the ground that groups of white-crowned lapwings leaped up to chase and screech at them, worried about their nests on the ground.

We drove down from the plain, still excited by our adventure, and entered a stand of mature winter thorn trees. It looked like an English wood: mottled light filtering through the high branches, draped lianas, and soft brown grass underneath. And yet, there was a herd of zebra standing in the middle, elegantly gazing at us. Not too English!

They spooked and raced off through the trees, cantering hoof beats reverberating.

And I finally learned why zebra all look to be pregnant, with bloated bellies: it's due to the gas produced by bacteria destroying the cellulose in their diet. Baby zebra didn't have the requisite bacteria in their guts and so have to be coprophagic (poo-eating) for a while in order to obtain them. Interesting.

We stopped by the riverbank and recognized this as the spot where we'd seen a flock of carmine bee-eaters two years previously. We reveled in the sight – and sound – of hundreds of white-fronted bee-eaters perching, flitting, calling with their little whiny complaint, and wiggling in and out of their nest holes in the bank of the river. There were also countless white-crowned lapwings in their brilliant white and black with intense yellow wattles and sage green legs, standing along the edge of the river bank. We had a tea break, and then drove back to camp through the winter thorn woodland (the pioneer species in the sandy soil) with one giant mahogany, the trees that will eventually take over in the next 50 years. We saw two elie bulls, one of them with six cattle egrets dancing about his feet, and watched them pull up clumps of grass and shake the dirt free.

These solo game drives that we'd exclusively enjoyed during our trip so far had allowed for fantastic personal attention and we'd loved learning so much about the wilderness of the African bush. In the area of Old Mondoro in particular, we'd not come across any other vehicles at all. It was delightful to have the chance to be so remote.

We hurried back to our room so that Jim could try again to repair the printer so that he could take photos of the staff, but was again unsuccessful. It was quite agreeable to sit on our deck overlooking the channel, with arrowmarked babblers and Meve's starlings literally at my feet. I could hear water dikkops with their "dead battery" call, and hippos chortling and woodpeckers tapping. Egrets lined the banks of the river, hunting.

We went back over to the main tent to take a going-away photo of Helen and Roelof with the staff, all dressed in their best kit. I was crying, and of course so was Jim, and so were Helen and Roelof. It was a very sad moment. Helen stepped over to the lunch table to cut the bread while we all cried even more. She was still, as always, so gracious. She looked up at me, starting to say "I'm ..."

I held up my hand, in imitation of Helen a few nights ago in our tent, and said, "If you're going to say you're sorry, I'm not going to accept it." She has taught me so much about hospitality. REAL hospitality.

We managed to eat some of the once again marvelously delicious lunch, and drank freely from the champagne. We had a very tearful goodbye to Helen and Roelof, and then we retired to our chalet to let them be with their staff for leave-taking. From their home.

Oh, I'm crying again now. They simply ARE Old Mondoro.

Anyway, I've got to pull myself together and keep writing. We had a fantastic siesta, secluded in our grass chalet. I read on the porch while Jim rested, and was pleased to identify a new life bird: the tawny-flanked prinia. Two little wren-like birds, each making little clicking sounds, pulled worms from the bush branches where they were hopping about, then perched, banged the worms against the branch (removing spines?) before sucking them into their beaks. In the lagoon, a black, brown and blue jacana was picking his way through the hyacinth, and our very own black crake arrived, delighting me with his bobbing tail while he explored the marsh.

We then took delicious hot showers, bliss in the bush.

Over at the main tent we had tea and snacks - chocolate cake and fruit pizza. Katie was doing well, overseeing the tea and making sure we had all that we needed. We went off for an afternoon boat ride with Andrew as driver, and had a fine time relaxing on the river and watching the afternoon sun on the tall green grasses with their puffy white tops lining the Zambezi. Just around the corner from our chalet I finally saw the water dikkops I'd been listening to during our siestas. We watched an evil-looking crocodile slide ominously into the water, and a monitor lizard sunning on a partially submerged branch.

In a little side lagoon, five buffalo splashed through the hyacinth, each with a single egret stiffly standing on his back. There were numerous jacana, showing off the bright blue lining the tops of their beaks in the sunlight. I really liked these birds, as Andrew reminded us that once the female laid her eggs, she left the rest of the child-rearing to the male only to go off and mate with someone else, starting the whole cycle again! Shocking behavior.

We saw all the "usual suspects" on the grassy islands and along the shore: hadeda, sacred, and even glossy ibis, egrets, guinea fowl - fighting and leaping around each other, goliath heron taking off and landing with gigantic expansive wings. A group of about twenty open-billed storks suddenly lifted off in formation just as we drifted by, causing me to catch my breath.

Just at sunset we pulled up to shore, meeting the new guests who have actually been coming to Chiawa longer than we have, and are great friends of the Cumings: Adrian and his son Richard, from the UK. We relished our Amarula and the marvelous orange sky, with the sun setting over the escarpment.

Our night drive back to camp was uneventful but splendid nevertheless; dazzling stars overhead, tiny flitting fireflies, and plenty of evocative night sounds whenever we stopped to listen. However, no predator simba or kaingo calls ... no alarm calls from prey.

Genet, civet, genet, white-tailed mongoose, genet, giant eagle owl. A lovely evening!

Back to camp with the usual lantern light, sherry, drinks by the river, snacks ... and a delectable dinner by candlelight on the riverbank, stars overhead, and wonderful conversation.

Butternut squash soup
Chicken (or veggie and cheese) kabob
Veggies
Potatoes
Lemon meringue pie

Richard offered us the use of his Sony video camera adaptor, so that we would be able to charge our video camera battery, and in the morning we could video tape the cacophonous hippo sounds and dawn chorus that we'd been thrilling to each day.

Off to bed listening to the cricket sounds in the bush.

And of course the next morning, when Jim had gotten up in the dark to set up the video camera, the hippos were literally silent. How do they KNOW these things! But we did get to record the dawn chorus of birds and the wheelbarrow bringing hot water for our sink.

After coffee and toast, we went on a game drive with a family from Australia (I still didn't feel well enough to go walking). We saw a group of vervet monkeys up in a large winter thorn tree, including some adorable tiny babies, hopping from branch to branch. We marveled at seeing a large baboon sitting atop a termite mound, apparently without a care in the world.

I was getting slightly bothered by the noisiness of the children on board; they weren't really misbehaving but nevertheless their boisterous fidgeting was bugging me. And then I saw a fork-tailed drongo, and remembered Kalpana's infinite patience with children. And I felt that I needed to be mentally kinder.

And then … in a grassy dambo … we saw …

A BLACK EGRET!

I decided I was being rewarded for my earlier personal reflection and appropriate improvement in attitude.

This little bird was simply standing in the wet green grass, gazing about, but it was still a thrill for us. He is an enthralling bird, and even though he wasn't fishing at the time, which is his neatest attribute, we were still energized.

We drove through the deep liana-covered forest, draperies of "balsam pear" vine covered with little yellow flowers. It looked really medieval and haunting; a mammoth gymnogene silently glided by overhead, his enormous gray wings slicing through the air.

We had an enjoyable experience next, by a bright green dambo filled with purple flowers. There were about thirteen male waterbuck – a bachelor herd – and we watched as they ate the hyacinth and marshy grasses, lit by dappled sunlight and just a few puffy clouds.

An incongruously outsized hammerkop nest was nearby, and little wartlets were running about. Stopping for tea and cookies, we had another debate about the famous squacco versus Madagascar heron, and think the two at this pond were also squaccos; we found some nice large white feathers for our hats.

Back in the jeep we listened to a squabble between two rollers with their grating voices, and watched a baboon baby proudly riding horse back on his mom. In another spot we came across at least a hundred red-billed hornbills, all silently flying away at our approach. I finally got a good sighting of the green wood hoopoe that we'd been hearing but had not seen. They are also known as "kilaga bafazi," which means "cackling women," as they sit all in a row on a branch, waving their long black and white spotted tails back and forth, jabbering like village women.

Retracing our steps back to the first dambo we visited in the morning to see if we could find a glove that one of the children had lost, we returned to the dambo of the black egret. We didn't mind the backtracking one bit as… YES… he was now fishing! He would run for a bit through the shallow water, then suddenly hunch down and spread his wings around him in a complete circle like an umbrella. Then he'd poke his head up, perhaps decide to pull his wings back in and run a little further before opening to an umbrella again … or he'd walk a few steps with a full umbrella and then dive his head down inside again. Meanwhile, he was wiggling his bright yellow toes, simulating worms under an overhang, so the fishes would come inside the darkness … and get nabbed!

I felt incredibly lucky to be here, once again experiencing the sublimity of nature.

On our way back to camp, we passed four bull elies – coming to say "goodbye" to us? Yes, clearly, as there were two more in camp that we watched while eating our delicious brunch, and then even more poignant was the elephant who walked right behind our chalet – almost close enough to touch – while we were packing! He had some marshy grasses hanging on one of his tusks (wearing jewelry for the occasion), and after passing our room he then proceeded to majestically swim across the Zambezi River.

What a grand finale!

We signed the book, crying when I read what Helen and Roelof had written the previous day, and then hopped on the boat for our one hour ride up river to CLZ. As we cruised, we saw hippos everywhere, bobbing down in the water (like the arcade game "whack-a-mole"), lazing on the grassy islands, sliding into the water from the river banks. For some reason I got the song "Nights in White Satin," in my head, and I wrote a song to go with the tune:

Nights with wild hippos,
guffawing without end;
lions and leopards,
in territories they defend.

Watching elies in woodlands,
always asking for more;
black egret umbrella fishing,
lapwings leap from the shore.
And I love Africa ...
Yes I love Africa!

I sang the song for Jim and he wasn't exactly moved, as he made gagging vomiting motions with his finger, in fact.

We arrived at CLZ to a warm welcome from Anna and Adrian. We were shown to one of the guest tents (there are four that line up along the Zambezi river), and then settled in the main gathering room for a discussion of our plans. The buildings include this main room, a large space with dark wooden floors, open walls, and a thatched roof; a similarly open office building; guest bathrooms; a chalet where Adrian and Anna live; the guest tents; a kitchen; staff quarters; equipment yard; and the Environmental Education Centre. The EEC has several school rooms and dormitories for the kids. While all of the buildings are connected with stone walkways, they are otherwise just simply out in the bush of Africa, red sandy earth and grassy woodlands all around.

The organization has three major goals: overseeing anti-poaching units in the LZNP; operating the EEC; and running a safari guide training school. They have been provided with seed money from the Danish government, but the grant ends at the end of this year and they would dearly love contributions to keep them going after that (to donate please visit: conservationlowerzambezi.com.za). We had met Anna and Adrian last year, had visited the operation, and were very impressed with how they were running it all on the limited resources that they had available to them. So, I hope that lots of people are moved to contribute.

Anyway, we wanted to help as we are passionately concerned about conservation of animals and believe that CLZ's three-pronged approach is brilliant. So, in addition to providing them with a solar panel for battery charging their anti-poaching patrol radios, and bringing a second one to give them on this trip, we had asked if we could come for a few days and volunteer our time. Of course they didn't dissuade us, and we gave a significant contribution for the privilege (they had to house us and feed us, after all). But, when we arrived and realized exactly how much work goes into running this incredibly diverse set of operations that is essentially accomplished by two people, we were a little overwhelmed. I'm not sure that a few days of our time were all that useful. Anna confirmed that it would have been more helpful if we'd come for a longer time so that we could actually finish projects. However, since we were here, she wanted to make the best of our time so she asked Helen, a volunteer visiting for the whole summer from the UK, to assign us some tasks.

One of the first things Helen suggested was that I help her clean up an EEC classroom, as the next group of village children would be arriving the following week and the classroom was looking a little shabby. We went to the room to remove all the posters and placards off the walls (which explain all about the different kinds of animals in the bush and about poaching versus protection) and then tried to wash the walls, to remove the old sticky glue tack. Whilst we were busy thinking of ways to deal with a leaky bucket (hang it out the window; use a stick to keep it from falling), one of the staff members brought me a little visitor.

It was Zamma, an orphaned baby elephant. CLZ had been taking care of this elie since he was found abandoned in the bush a few weeks ago. And I fell in love with him! He was very tiny - for an elephant - standing about 2 ½ feet tall, weighing perhaps 300 pounds, and covered with very fine hair on leathery skin. They thought he was only about four months old. I didn't realize how difficult it would be to "cuddle" with him, as even as a baby, he was relatively large, rambunctious and unpredictable. He is, after all, a wild animal, and when he head-butted, he could really break bones! Even so, he was adorable. As he was teething, just like a human infant, he loved grabbing my arm or my hair with his trunk, and pulling them into his mouth. And even though he was rather grimy and left a lot of smudges all over my clothes, I still loved petting him and scratching him behind the ears, and giving him little hugs.

Zamma needed to be fed milk (of a very particular formula, developed by Dame Daphne Sheldrick in Kenya over the past 30 years; check out their wonderful orphan elephant organization at www.sheldrickwildlifetrust.org), from a bottle, every two hours around the clock, for two years! Wow, that's a huge commitment. So after we realized that washing the walls of the EEC wasn't going to work, Helen suggested that I assist in weighing out the powdered formula that Zamma would consume over the next 24 hours, and sterilizing the containers. He recently had a bacterial infection so they were trying to sterilize things as best as they could. Water was boiled over a fire and used to wash the tins before drying with clean towels and adding the powdered baby formula. There were two Zambian "keepers" who spend all their time with Zamma, feeding and watching over him, and it was interesting to see that they didn't really understand the steps involved in sterilizing baby's bottles; after all, in their villages no bottles were used, babies were only breast fed. I had some conversations with these staff members and while it was weird, it was sort of nice to be called "madam," as it seemed that this meant I was an elder in some way. It seemed very respectful and made me feel grown-up.

Helen and I walked out to the sandy hillside where they had made a mud wallow for Zamma to take his bath; baboons were walking about the sandy hillocks as well. Unfortunately Zamma didn't seem to want to get in the water - perhaps it was too cold as there was a slight wind, or maybe he needed other elephants as company. From watching the large breeding herds it is clear that babies love getting in the water, so there had to be some reason Zamma was so uninterested.

Jim, meanwhile, was spending time with Adrian, driving through the bush in the safari jeep (yes, he even got to drive), looking for a young elephant reported to have a wire snare caught around his neck. When the elie was found, CLZ would arrange for a vet to come and dart him, to remove the snare. Unfortunately he wasn't found while we were visiting, but the search continued as all safari camps were informed to keep their eyes open for him.

When they returned, Jim came to the EEC classroom to see how we were doing, and the three of us came to the conclusion that the walls simply needed to be repainted. So, Helen arranged to get all the stuff we needed (buckets of paint, paintbrushes, and rollers) while we spent the afternoon removing the bits of sticky glue stuck to the walls. For this job we exhibited great creativity, finding machetes and picks from the classroom displays about poaching and farming to be particularly useful tools for the task.

After a long afternoon spent working up a sweat, when it got too dark to see we ventured over to the riverbank fire circle for drinks while Adrian cooked dinner. A beautiful orange sunset, a tiny sickle moon overhead, and Venus setting made for exquisite ambience. We were joined by several other dinner guests - Riccardo, a CLZ board member who was visiting from his safari lodge further down the river, his friend Sylvia, Helen's boyfriend Danny who was volunteering at Chongwe Camp, next door, for the summer, and Troy - also from Chongwe Camp. As we relaxed in this enjoyable company, potatoes and a stuffed butternut squash wrapped in foil cooked in the coals, steak and veggie kabobs grilled over the fire, and sausage cooked in a "pojke" pot placed into the fire. When we walked to the main room to consume this delectable meal, hot freshly baked rolls, salad, and more wine awaited us.

Dinner segued into quite an evening of hilarity chatting with this group. We learned all about the games that the safari employees play when they have a night off and hang out at CLZ, like seeing who could pick up a cereal box from the ground from a standing position, using nothing but the mouth, and then at each success, the box would be cut into smaller and smaller pieces;

seeing who could reach the furthest over a line in the sand, using no hands but allowing for two team members to assist in cantilevering (they told about one young lady "launching" over the riverbank during one of those attempts). The ribald sense of humor increased as the night progressed.

Eventually we retired to our room, a traditional safari style tent with zippered mesh screens on front and back, mounted on a concrete pad with a square roofless room of concrete walls out back with shower, sink, and loo. Quite comfortable! We pushed the twin beds together and since it was cold at night we managed to snuggle a bit.

Our tent looked right out at the Southern Cross suspended over the river, hyenas were giving their haunting "whoooop" calls as we drifted off to sleep, and we got to hear hippos – and lion calls – all night long.

I had offered to get up at 1:30am to do a spot check for Zamma's keepers, relieving Anna and Adrian from this job of helping to ensure that the keepers manage to wake up to feed Zamma every two hours. It was quite scary for me to be walking through the African bush alone in the dark of night, as elephants, lions, and leopards did come through this camp, after all. With only my little head lamp available, I carefully scanned the bushes on both sides of the walkway as I went. It was a rather long walk and I made my way quite slowly. When I got to the kitchen the keeper was already inside, washing the bottle after feeding. I still went over to the enclosure where Zamma was kept at night and gave him a pet as he was standing near the warm fire.

"Hello!" Adrian announced from the shadows. I guess he'd not known whether I would in fact remember to wake up, so he'd come anyway.

Lion called again early in the morning. I was glad he didn't call while I was out walking around as it may have led to dire consequences!

We were up at 7am, for a quick breakfast. Riccardo quipped, "I just thought I should mention I noticed Cynthia leaving her tent at 1:30 in the morning ... and walking towards the staff quarters!"

And then we were off to the EEC room for the painting prep. We sanded down the spots where the glue had been (that was hard on the fingers) and then Jim taught Helen and I good painting technique, as we painted all the edges, corners, and sanded spots of the classroom. It was really hard work. At one point I took a coffee break to chat with Adrian about his getting his pilot's license in Palo Alto (since they don't teach flying in tail draggers in South Africa). We offered that he could stay with us and borrow a car, and that should help with the cost of getting his license.

We had lunch with the safari guide trainees, and it was fun to eat the traditional Zambian maize called nshima, which was quite tasty, and beans and kale, although they didn't insist we eat the kapenta – small anchovy-like fish from Lake Kariba.

Our afternoon was spent with more painting, and I alphabetized the library. It felt good to recognize many of the books we'd donated last year. I had sent several boxes of my favorite stories for sharing with the staff and the EEC. Mid afternoon, Anna brought us warm cheese scones that she had just taught the cook to bake. Divinely delicious.

Later, Jim and Adrian asked me along for another search for the snared elie, driving up towards the escarpment along the Chongwe River. Of course I jumped at the opportunity to be of help, and as we drove I scoured the landscape for elephants. It was also quite interesting to chat with Adrian about running CLZ and some of the experiences he and Anna have had. Adrian recently took a "darting class" in Zimbabwe, and got his certification for administering tranquilizers and their reversers. It sounded fascinating. The students in the class darted lion, leopard, giraffe, and elephant for practice. The biochemistry of the opioid compounds was also intriguing ... and the fact that the tranquilizers could kill a person within minutes if they got splashed onto a cut or otherwise injected into the blood stream ... but if the reverser is handy the person could be saved.

We ended up driving to the top of the local foothills, and were treated to an awesome view over the LZNP. Adrian suggested that we hop out and clamber over the smooth red, pink, and gray rocks (I was loving this part!) to see the Chongwe waterfall, where he and Anna have enjoyed picnics on those rare occasions when they can take time off. It was wonderful that there was still water flowing in the falls, making it even more picturesque. I noted that it felt "interesting" to be walking around in the bush without a rifle.

Back in the jeep coming down the hill, I continued to scour the landscape for elephants. But of course, as we rounded a particular set of bushes, I had begun chatting about something and didn't even see the group of elephants standing right on the road until Adrian halted the vehicle. Great help I was, huh?

The matriarch of this group was tuskless, and perhaps that's why she appeared to be very ornery. We had a stand-off for quite some time, and she wouldn't give ground. I was impressed with Adrian's foresight for not turning off the jeep, just in case we needed to make a quick escape. It turned out that eventually we simply backed away, and drove home via a slightly different route.

We also heard some really good news today - the anti-poaching patrols had recovered some ivory, and, more importantly, the poachers were no longer at large.

Back in camp we had another scrumptious dinner with the gang, chatting about relationships and other life topics. I really love being friends with these people who are trying to make a difference in the world.

> *Spaghetti with either meat or veggie sauce*
> *Cauliflower and string beans*
> *Fresh baked rolls*
> *Chocolate cake*

Back in bed, cozy in our tent by the river, we listened to hippos and hyenas calling all night. At one point I awoke to hear a nearby elephant munching and breaking branches, and before drifting back to sleep I remember hoping that he'd not still be around when it was time for me to go visit Zamma.

However, when 1:30am rolled around and Jim's iPhone awoke me with its alarm, I could still hear the munching and branch cracking nearby. Nevertheless I got up and put on my contacts, and ventured out into the night. I was thinking that if the elie wasn't exactly on my route it would still be safe to go to Zamma's enclosure. I was really quite scared, however, and took only a few steps at a time before stopping and listening, straining to figure out where he was. It was pretty obvious, unfortunately, that the elephant was literally right in front of me, and I only got about two tent-lengths away before giving up and turning back. Just what CLZ didn't need was a trampled volunteer!

We re-set the alarm for 3:30am, and that worked out much better. I got up again, put the contacts back on, and managed to visit Zamma and his keeper, who was in the process of making the next bottle. Zamma was peacefully slumbering, lying on his side next to the fire, covered with a warm blanket. It was really an endearing sight.

The next morning we were up at 6:30 again, raring to go, as it was time to paint roll the walls of the EEC classroom! Jim did all the rolling, but I managed to be of some assistance by removing the old paint from the roller pan. In the meantime he was using glass lasagna pans, and they weren't working as well. The old paint was hard to chip off so it was quite helpful that Helen suggested using petrol to dissolve it which worked miracles. Jim then had a much more manageable job, and finished it in time to go with Adrian to Royal Airstrip to check out the Scout plane that belongs to CLZ, which could be used for anti-poaching when Adrian gets his license. They taxied around the runway and Jim said it was VERY tempting to simply apply full throttle and take off ... but luckily, since his license isn't current, there wasn't sufficient gas in the tank.

After spending some time with Helen and Anna in the office, and appreciating the CLZ T-shirt that Anna gave me, I realized it was time to gather our luggage from our tent.

Easier said than done!

An immense bull elephant named "Oliver" (by Anna, who pointed out that his tail had a "TWIST"), had arrived on the scene. He was in front of the row of tents along the riverbank, chewing on the bushes. Anna and I snuck around the back of the main gathering room so that I could try to get to our tent; Helen wanted some good photos so slithered under the decking of the room until she had a good view out the front. I was able to sneak along the backsides of the tents, and slip in between the concrete wall of the loo and the back of our tent, unzipping the back flaps to get inside.

I was planning to grab our luggage and escape, but at that moment Oliver arrived out in front of our tent. At first I excitedly filmed him a bit through the front flaps. Then, when I started to organize our belongings, he suddenly became interested in what I was doing. I was surprised that he seemed to be paying so much attention to me, and a few times when I needed to go near the front of the tent, he REALLY paid attention. He turned directly towards me, taking a few steps in my direction, and flared his ears slightly to make his point. One time he got so close that I was really frightened – my heart pounded as I clumsily leapt backwards onto the bed and repeated under my breath, "they don't come into structures; they don't come into structures," hoping that this was really true. And apparently, once again, it was, as he didn't venture into the tent.

After being trapped for some time, I heard Jim calling from the main buildings. I replied that I needed a little help, and he snuck behind all the tents to the backside of ours. I handed each of the pieces of luggage to him, while Oliver seemed to be looking the other way, and we made our escape. That had been really quite exciting.

We had a teary goodbye to Anna and Adrian, Helen and Danny, and climbed into the boat from Chiawa. Although it was hard to leave CLZ, we simply adore Chiawa Camp and were elated about getting to stay there once again. When we arrived, Lynsey and the new camp hostess, Connie, met us and offered us a drink in the main buildings. We love these spacious structures set on smooth concrete, with their thatched walls at the back and completely open fronts leading down the sloping lawn to the Zambezi River. Dotted with comfortable chairs, bookshelves, and a full bar, and with an upper floor for wildlife viewing or star gazing, the central area of the camp is comfortable and inviting.

Lynsey then walked us to our chalet … and surprised us with an upgrade to the honeymoon tent! I guess I don't need to mention that I cried again.

Picture a large safari tent mounted on a smooth dark wooden platform. The loo out back has dark wooden walls and an open ceiling which allows me to indulge a favorite activity – birding in the loo! Twin sinks and a claw foot tub with candles and bath potions make for the perfect bathroom – just splendid! We were also pleased with the location of the room, as it is the tent furthest out on the point of the confluence of the Chiawa and Zambezi

Rivers, and as such looks right out on the grassy flood plain where baboons, impala, warthogs, and even elephants are often seen.

We started with a superb outdoor shower, and took photos of each other sitting on our deck in front of the room - with either an elephant or a buffalo striding right behind us.

We met for tea and snacks at 3:30pm, and spent some time in the little gift shop, picking out Chiawa shirts to bring home to Ellen and Ron. We'd had a superb time together at this camp last year, and we wanted to commemorate that trip.

We were then off for our afternoon game drive with Adrian and Richard on the jeep, the same guys we'd spent time with at Old Mondoro. Boaz (the uncle of Dispencer, a guide we've spent time with at Chiawa before), was our guide while Wallace spotted. We mentioned that we love birds, and we stopped for many, some that we hadn't seen on the trip yet. Our sightings were now up to 132 for the trip including 14 lifers.

We came across lion spoor, and followed it as best we could, considering that there were only intermittent dirt areas and we couldn't easily track through the grass. We searched in and out of thickets, bushwhacking bravely but no luck. We did see a herd of hundreds of buffalo amongst the trees, and were happy to see many revered baobabs.

I was unhappy however, to notice that a tsetse fly bit my ring finger, and it immediately began itching. Luckily, I was prescient enough to take off my wedding ring, as during the drive my finger swelled up quite impressively. I asked again about the likelihood of finding sausage fruit and Boaz promised to keep his eyes peeled but didn't know of any trees already bearing fruit this early in the season.

Our Amarula sundowners were served on an open red sandy plain, with the crescent moon and Venus hovering over the dark black escarpment sharply outlined against the orange red sky. The scene filled my soul with a peaceful joy.

We began our night drive by inadvertently passing a breeding herd of elephants immersed in some nearby thickets and were treated to a really loud trumpet warning, a sound which evokes a primeval thrill.

We saw two baboons bedded down for the night in amongst the leaves of a towering palm tree. It was hard to imagine that this could be comfortable considering how razor-sharp the edges of a palm leaf are, but perhaps it was worth a little discomfort to be safe from predators.

Later in the drive I was sure we heard baboons barking – juvenile baboons, perhaps, as it wasn't as strident as the adult calls we've heard. But Boaz politely informed me that it was genets fighting. We've never heard a genet make any kind of sound before, so that was interesting.

Bush hare. Genet. Genet. White-tailed mongoose.

We came across a bull elie reaching up into the high branches of a winter thorn tree, and Boaz guessed – correctly – that he was going to rear back, further and further, and then he STOOD on his back legs, front legs leaving the ground, to reach even higher into the tree. It was captivating to see such an immense bulk lifted up in the air! Boaz worked really hard with the spot light, to be sure that Jim got this cool maneuver on video.

A Scops owl was next sighted, and as this was a bird we'd never seen but had listened to for years, all night long in some cases, it was really satisfying for us to see. And it is a particularly cute little owl.

Three different side-striped jackal couples were running along the sand, just before we pulled up at camp where we were received with welcoming warm towels and glasses of wine at the bar.

And then.

One of my favorite moments in all of our Africa travels.

The Chiawa choir came to us, marching two by two in time to their music, singing African songs and inviting us to dinner.

And now I really cried.

Dinner was served at long, elegantly set tables, complete with white tablecloths, china, and candlelight, on the grassy bank of the river, under the slowly turning wheel of stars. We merrily chatted with Grant, while I enjoyed my favorite dish in the world - gem squash - a round summer squash with a very hard outer shell, scooped out after baking and mixed with luscious spices before returning to the shell.

> *Fried halumi (cheese) with jam and almonds*
> *Butternut squash soup with curry*
> *Pork or lamb (gem squash for me)*
> *Rice*
> *Veggies*
> *Apple cobbler in phyllo with caramel sauce*

We then got on the "taxi" to our room, at least that's what we affectionately called it when the staff member with the torch accompanied us down the path behind the chalets, and stopped to drop off each couple at their tent.

Our room was lit by kerosene lanterns and candles, and the mosquito netting had been romantically draped over our comfortable bed. There were chocolates on the bedside table, and hot water bottles covered with soft fuzzy covers nestled amongst our fluffy down comforters. Simply blissful!

During the night, I heard a hippo walk by the tent, and laughed at the splatting sound of his spreading his dung around by energetically twirling his tail, aka the famous "shit helicopter," whereby he marks his territory. There were also a multitude of thunderous hippo calls (yippee!), although that might have simply been because the video camera was being charged so we had no chance to record the sound. I was able to enjoy these sounds quite frequently, as I was kept awake much of the night with the throbbing of my tsetse-bitten swollen ring finger. I wrapped it in a cool washcloth (no ice cubes available out here) and kept it elevated, but it was still painful enough to interrupt sleep.

But happily, I also heard lion calls during the night. Nearby. Truly the most redolent of African night sounds.

We were awoken with a "Good morning," as tea and coffee were delivered to our deck, and I hopped up to serve Jim. We hustled to dress and were soon out at the fire circle by the river, eating Amarula-flavored porridge, strawberry/white chocolate muffins, and drenching ourselves in the beauty of the dazzling red sunrise.

This morning I finally felt recovered enough from my tummy incident to go for a walking safari, and we were delighted that Grant offered to be the guide. We went by boat to a spot several miles downriver, and walked all the way back to camp over 3 ½ hours. It was an excellent opportunity to finally get some exercise – I was actually tired out!

Being on foot in the wilderness of Africa is always an adventure, silently stalking, single file, following Grant and a scout with a rifle. I was especially alert to sounds on all sides, realizing that at any moment we could come across a large predator. Made my heart pound!

We stopped every once in a while, so that Grant could point out an appealing bird, or a remarkable plant, or to discuss the ubiquitous dung and termites.

We came across a quiet glade with a small dambo. An imposing buffalo stood in the water, guardedly staring at us. Grant fittingly quoted Robert Rourke, "He has a murderous look in his eye."

I liked Grant's description of the impala middens, where each male deposits his scat pellets on top of the other male's scat to prove his importance, as "poo wars." He said that since they couldn't build walls, or place rings on their sweetie's hooves, this is their only way of proving their dominance.

Jim asked what HE was supposed to do, now that his sweetie couldn't fit her ring on her finger? My hand, by the way, was in agony during the walk, puffed up like a glove.

We stopped by a particularly lofty termite mound, and Grant asked whether we'd all heard "The Termite Lecture." There were three other people on the walk with us - a South African couple and Lynsey's sister Laura, and we had indeed heard various termite lectures over the years, but we wanted to hear it all again. As always, it was incredibly fascinating …

Termites manage to keep the temperature inside the mound almost constant at all times of year, due to the convective effect of rising cool air from the subterranean tunnels, as they alter the amount of venting on the sides of the mound appropriately. The queen and king lie in the royal chamber, totally monogamous, for seven to 12 years. The queen is a slug who can grow to be five inches long. Eeeuw! She's fed and licked (eeeuw again) by the worker termites, to stimulate her to lay eggs constantly. These workers leave the mound at night to eat vegetation, then come back inside and deposit their excrement on a fungus farm that they tend. The fungus breaks down the cellulose fibers for the termites and then they can re-eat the dung. Soldiers are much bigger and have larger mandibles; they leave the mound to protect it and might even get sealed outside if the threat requires damage repair. The winged forms, male and female, are the only ones that are fertile. They live under the fungus and come outside only for their "prenuptial flight" ... they land and shed their wings and secrete pheromones until they pair up. At this point, most get eaten (by animals, including humans), but those that survive as a couple can start laying eggs to produce their own workers and soldiers, using the fungus spores on their backs to start a fungus farm. In seven years the new mound would be the size of a soccer ball. So imagine how many years it had taken to make this mound, which was about 15 feet high!

The soil of the mound was also quite fertile; not only used for the walls of houses in Zambia but pregnant women eat it, and plants happily grow in it even if the ground surrounding the mound is dried-up and infertile.

We were also quite fortunate to see a family of elies, although at a distance, along with some waterbuck, including babies, and warthogs. Near the camp we came quite close to a giant bull elephant giving himself a sand bath. He didn't notice us at first, and so Grant angled our walk so that we wouldn't box him in between ourselves and the river. Even still, when he did get wind of us he gave a very nice threat display with his ears full out, staring us down.

Lunch in the dining chalet by the riverside was delectable with many courses, and while we were eating Grant and Lynsey brought over my "secret mission" surprise for Jim: six carved wooden guinea fowl like the ones we love at Sausage Tree camp. We BOTH cried! It was a special, sweet moment, and made even more poignant by the card they gave us, explaining that these were a gift from them to us.

Back in our chalet we took a delightful bubble bath, and gazed out the large open window to the river. There was a family of elephants in the sandy riverbed of the Chiawa, slowly walking towards camp.

I ambled over to the main lodge and played with Scott for a while. I had offered to baby sit him so that Grant and Lynsey could have a little break, but

I'm glad Lynsey stayed by my side as the bush is a veritable mine field for a toddler with energy!

Grant came by to chat with me about our passports, wanting to make sure we had our visas properly taken care of, as there had been some trouble with this in Zambia recently. And that's when we learned that we'd actually overstayed our visa - a scary pronouncement, for sure - as they'd erroneously noted that we were only going to be in the country for 3 days (remember the conversations we'd had with immigration when we entered into Zambia?). Yikes! Dave, Grant's father, then made several trips to immigration for us, and along with phone calls from Jenny, we were released from the possibility of a trip to jail or a $250 fine. The Cumings really protected us.

I walked back towards our chalet to meet Jim, and we were entertained by a large group of vervet monkeys in the trees alongside the trail, leaping and racing along the branches, boxing and hopping about. Right behind them was a bull elephant, contentedly eating some of the branches of the fence around the kitchen enclosure.

By tea time I couldn't handle the tsetse bite any longer, and finally took an antihistamine and Sudafed. Earlier in the day I had taken ibuprofen but it had had no clear effect on the swelling or pain. I had also tried an antihistamine cream that Lynsey provided, even slathering it on thickly all over the hand, but this didn't seem to have any effect either. So now I had a swollen, itchy, painful hand AND I was drowsy. Harrumph. I needed a sausage fruit! Apparently most people don't have such a strong reaction to tsetse bites, but for those of us who do, the sausage fruit is imperative.

After a tasty snack, we hopped in our jeep for an afternoon game drive, with Temmy and John, a very nice couple from Santa Barbara. We asked Boaz to find some elephants, and immediately as we left camp, there they were. Ask and you shall receive? Four elephants right IN the road, two youngsters pushing and shoving each other, fighting over the umbrella acacia pods they were picking up from the ground after a larger elephant had shaken the tree. We had a good time watching them, but eventually had to back up and go a different route, after several ear flaps informed us that they didn't intend to give way.

We encountered many different troops of baboons, relaxing in grassy dambos, many with babies in their arms or on their backs. When we got to the largest dambo, we were treated to a wonderful spectacle of elephants all around! We experienced several loud trumpets, and sat still, watching, while elephants paraded around us. I loved it.

After this lengthy encounter, we had to have "sungoners," but no complaints, as it had been certainly worth missing sunset. We drank our Amarula contentedly before leaving for our night drive.

And then, it happened.

Boaz had spotted a sausage tree, and there was a sausage fruit ON THE GROUND! Yeah! Wallace got it for me, sliced it in half with his panga, and I started to immediately rub it on my hand, digging out bits and pieces of the flesh with my fingernails. All my fingers ended up being stained brown but I didn't care, as the swelling thankfully went down. Each time I experience this I am freshly amazed. The itching was gone almost immediately, and within two hours my hand was no longer sore and almost back to a normal size. Miraculous relief. There is a company in South Africa which is investigating the molecular mechanism whereby the fruit assuages tsetse bite itching ... and they may have a sausage fruit salve available for sale soon.

And then, on our drive, we saw:

Civet.

Porcupine - from a large distance, unfortunately, and just for a few seconds before disappearing into the brush.

Genet.

Honey badger – also from a large distance, unfortunately, and just for a few seconds before disappearing into the brush.

White-tailed mongoose.

And then, we heard lion. Yes!

Boaz debated ... the call was from quite far away ... and then we all decided to go for it. We raced through the night, our hearts pounding ... and then stopped and listened.

Waiting quietly. Every sense alert. Listening intently.

Simba!

We continued the chase, now on extremely bumpy ground, driving on and on.

Stopped again, listening and focusing. A long time.

Nothing.

We gave up, then, and started home, but then heard another call, this time VERY close! Perhaps within 50 meters! So we drove and drove and drove over the insanely bumpy ground of the dried-up dambo.

But never saw him.

That elusive Douglas.

We were back in camp quite late, but again had the opportunity to cry while listening to the Chiawa choir, and then to consume another mouth-watering meal out under the stars.

Couscous with prawns (mushrooms for me)
Vegetable cheese soup
Beef or bream
Potato croquette
Veggies
Chocolate mousse over crumbled cookies

We had a glorious night in our comfy tent (although I was zorked by the Sudafed), with lots of lion calls. There were fewer hippo calls in the morning, of course, since Jim had set up the video camera.

This morning we had whiskey-flavored porridge and raisin bread, then zipped off for another game drive with Temmy and John. Dispencer was our guide and again, I just barely managed to NOT call him "distemper" by mistake. I told him that while everyone else might be interested in finding the lion who'd been calling all night, I was much more interested in finding a guinea fowl feather, as mine had fallen out of my hat. He laughed but at one point during our drive he stopped the jeep and stepped out in front, presenting me a guinea fowl feather with a flourish!

At the beginning of the drive, I spotted a lion spoor. A very LARGE lion spoor. This was clearly Douglas. We followed the prints for quite some distance. We came across a dambo with two exquisite saddle-billed stork (male and female; one with black eyes and one with yellow), strutting about in the hyacinth. A gray heron posed nearby, until a fish eagle dove on him. The eagle landed to perch in a nearby tree and we could watch him as he screeched his haunting call, flinging his head back each time he called.

We saw more Douglas spoor and followed it, and followed it, and followed it. I loved seeing the regal, dainty crowned lapwings with their bright orange legs and distinct black and white crowns, and thought about how I preferred these royal-looking birds to the constantly yapping white-crowned lapwings that are so much more prevalent.

We were still cat searching - "looking for pussies," as Dispencer called it, but stopped to take our photos in front of a gargantuan baobab before a tranquil tea break. Jim found a miniature preying mantis and harvester termites, the only termites which can come out in the sunlight.

Back on the road again, we passed several bull elephants, and then watched a mom and baby. We saw her shake a palm tree, and could see other nearby trees trembling as well. There must have been more elephants in the foliage.

At another dambo we saw a tiny baby hippo next to some big guys, along with egrets and an elephant stopping for a drink. Several sets of impala were locking horns and jousting; a scimitar-bill wood hoopoe perched overhead with long black tail and white spots, easily differentiated from the green wood hoopoe as it was solitary rather than in a group of cackling women.

Two endearing baby wartlets allowed us to get a very close look - normally warthogs run from jeeps but these were too young to be capable of living alone in the bush; their parents may have been killed. It is doubtful that they would last long alone at that vulnerable age. They are really lovable looking, with fluffy manes reaching all down their backs.

And then, we experienced an elephant ecstasy. We pulled up to the Zambezi riverbank, and found ourselves in the middle of a group of elies: a tremendously large bull, giving himself a sand bath; another large bull, resting his trunk on his tusk; two younger elephants at the river's edge, drinking in concert; a mom with two youngsters and one tiny baby, nervously assessing our behavior. And while we watched, the wind and waves of the Zambezi were lapping at the shore.

Dispencer had his hand on the ignition the whole time, so when the largest bull suddenly turned on us and flared his colossal ears, taking a step towards us, he was ready to pull the jeep back and move away. It's great to know we are well protected in the bush, by the Zambian guide's depth of knowledge.

We had another yummy lunch back in camp, and then a luxurious bubble bath while Jim washed my hair. After a short siesta we met our guide, Laxton, in the main building to be driven upriver for an afternoon canoe ride. As we motored along the river, we were treated to an extensive breeding herd of elephants at the water's edge. Cruising along in a boat we got closer to them than we would have in a jeep. I was enthralled, especially watching the babies as they tried hard to properly maneuver their unwieldy trunks.

We continued upriver to the "Grand Canyon of the Zambezi," a spot on the riverbank where there are deeply chiseled, eroded cliffs reminiscent of buttes in Arizona. Here we got our excellent briefing for the trip from Laxton, who explained all that we needed to know. We had asked that we be treated like king and queen this time; instead of paddling the canoe ourselves, we wanted to have the guide do the work so we could take photos and video. Laxton was in the front canoe with another guest, and Wallace was in the back of our canoe to do the work, although Jim was willing to pitch in now and then.

We spotted a young monitor lizard on the river bank, and learned that they lay their eggs in termite mounds … as we had learned, the temperature inside was so well regulated that it is a perfect incubator!

Leaving the Zambezi River we entered a beautifully tranquil side channel, and laughed at the rather comical sight of two hippos standing on the grassy verge staring at us, then each in turn literally pronking before waddling to the edge and making a leap into the deep water right in front of us. Zebra stood on the cliffs ahead of us, hesitating before galloping away, and then, after we had entered the channel we were able to look back and watch a whole breeding herd of elephants crossing the water behind us. A rapturous moment, watching their splashing and wallowing, trunks held high.

We had a fabulous canoe trip down the channel, with little bee-eaters flitting about, gray louries plaintively calling, gigantic evil crocodiles sunning on the banks, baboons grooming each other, waterbuck running and splashing through the grasses, vervet monkeys in the trees. As requested, we arrived back in camp for sundowners out at the point, in time to take sunset photos with the ideal backdrop of Chirapira Mountain. Lisa, the hostess, was there with an elegantly set table – it was actually a mokorro, an old-fashioned boat carved from a tree, cleverly mounted at table height and draped with white linen, and our Amarula and snacks - sausage rolls including veggie sausage rolls for me.

Then, we were off for our last night drive (no dry eyes for Cyn), with Mike as guide and Laxton as spotter. Right out of camp, we saw a group of impala all bunched together, acting terribly skittish - a good indicator of a nearby predator. Then, we heard a solitary bushbuck warning call. And we were off! We zoomed around, looking, up on the hills, down in the bushes.

When we had hopped in the jeep earlier, I had mentioned that I liked nightjars, and as we were frantically searching for predators a nightjar flew right in front of us, and simply wouldn't leave us alone, flying alongside. Mike joked about whether or not we should stop to examine it for my sake. NO! Keep searching for leopards!

But nothing.

Hmmm.

We eventually decided that the impala and bushbuck had just been anxious, because it was the time of day where the light is changing and it is difficult for prey animals as their vision is compromised.

So we gave up on this focused search and continued along the track to see what we would see.

And we saw nothing.

Really! No animals of any kind. At this point, not even a nightjar!

At one point it seemed time to head back home, and I exclaimed "not even a genet" …

And then we saw a genet!

So, satisfied with our full night drive (ha ha), we did start home. … and came across a bushbaby. Two of them were clambering about in the tree and we got a good look at them. With their long thick tails, they are imaginatively named "thick-tailed bushbabies." Mike explained that they are territorial, marking up to 1,000 trees in a night with their sticky feet.

After this, we were back on the main road (still, just a dirt road), hearing from Boaz via radio that their vehicle similarly hadn't seen any animals in the night, and Grant called to say that he'd overheard some other radio calls saying that the only thing out and about was something on the main road.

And then.

Just when we came around a thicket, right in the middle of the main road.

DOUGLAS!

SO regal. Really the king of the beasts. He was sauntering along slowly, sniffing the air, with head held high. He is truly a gorgeous lion, with his taut sinewy muscular body, lavishly flowing dark mane, purposefully striding through the bush.

So much more beautiful than the scruffy, younger lions further east.

Douglas climbed up on the sandy hills and then flopped down for a rest; we watched him for quite some time but when he got up again and eventually walked off to the bushes, we left him in peace.

As Mike explained to us, it's really about time that one of the other lions usurped Douglas's reign, since his daughters would soon be old enough to come into estrus, and it was therefore appropriate for another male to take over the harem.

We hopped out of the jeep a little later to examine some Matabele ants marching in formidable formation and copped an earful of their quite loud hissing when we disturbed their ranks.

We then drove down into the Chiawa riverbed - for a barbeque in the bush! This tradition of Chiawa's is a magical experience ... kerosene lanterns set in the cliff walls, long tables in the sand elegantly appointed with china and crystal, bonfires burning all around. The choir sang, I cried, we enjoyed dish after dish of delectable food.

Douglas called, and I cried again.

We were driven back home late at night, happy and full of good food and wine, and listened to hyena calls. Another delirious night, with hippos burbling and fish eagles wailing.

In the early morning, I heard a lion call nearby, and so after coffee and tea in our chalet, and breakfast by the river (rum/raisin porridge, banana bread), we hopped on the jeep for our last game drive of the trip.

A trumpeter hornbill called overhead, crying, like me.

We looked for Douglas's prints, found them, and began tracking.

"I love the search for lions in the morning," John exclaimed.

And then, Boaz drove us right up to ... Simba!

Douglas and his family - all lazing about in the red sand. Three young cubs a little over two years old (perhaps the same ones we've seen the past two years), an adult female, and a subadult male. A lovely grouping.

We watched them sleeping for a while, then continued on, sunshine lighting up the shaving-brush combretum in the crisp air.

Warthogs running with tails pointed straight up, redbilled hornbills hopping and scrabbling for termites in their little mounds, a lappet-faced vulture perched on high.

Boaz found a potato bush, as requested, and we collected seeds to bring back to Ellen as a special gift from Zambia. She loves that mashed-potato smell in the dusk!

We stopped for coffee by the twin baobabs, enjoying the tiny banana bread muffins and chatting with each other. Driving once again, we passed a fork-tailed drongo … Jim and I exchanged a meaningful glance.

And then, we saw a carmine bee-eater!

"Stop!" we begged. He was perched right on a tree snag, right next to the vehicle. Amazing colors, long racket tail. They don't migrate to this valley until September; Roelof had said there may only be ten birds or so that stay over the winter. Guess we were really lucky!

See, I knew the fork-tailed drongo was a sign. Patience pays off.

We passed a family of baboon, mom hugging child in a human-like embrace. We ducked back behind camp for a last look at Douglas and his family now relaxing in the shade. It was so nice that we got to see him again this year, as you never know…

A fish eagle lifted off from a branch right overhead and then flew away … as must we.

Back at our chalet we took a final outdoor shower, packed up our meager belongings, and then enjoyed another of Chiawa's signature pleasures: champagne brunch on the river launch. As always, we were elegantly served a complete scrumptious meal, while quietly motoring around the river and its islands, and had an amiable time chatting with Lynsey and her sister. The whole Cumings family is truly wonderful: not only did we experience Lynsey's hospitality, Grant's brilliant guiding, and Dave's fascinating company and multiple extended trips to fix our visa situation, but Jenny is a similarly magnificent person. She offered that we could spend our last night at her home in Lusaka where she took us shopping for gifts. She also took us out for a wonderful companionable dinner where we met another famous couple of camp owners, Chris and Charlotte McBride, who have been guiding from "McBride's Camp" in the Kafue National Park for decades. If all that wasn't enough, at the airport early the next morning, Jenny's special help facilitated an upgrade to nicer seats and assisted us yet again with tiresome customs and visa issues.

With one hour remaining before it was time to leave, we went to visit Temmy and John in their tent, as they wanted to show us how nice it was. We've decided that next time we will request this tent, number nine. It is fantastic, with the shower literally IN a tree, and a wide open loo, and smooth large deck out front, and immense king-sized bed.

As we were leaving, we noticed a few elephants down by the grassy confluence of the two rivers, and so we wandered over to watch. We set up our little camera to take a photo of both of us, with the elephants in the background, and then, we had the most sublime moments of the trip!

More and more elephants began arriving, running to join the others, drinking in the river, thrashing about in the mud, and taking sand baths. Little tussles, a baby flinging himself in the mud face first. We heard rumbling and watched greeting ceremonies, and couldn't believe how many elephants kept arriving. We watched, completely enthralled.

At one point we thought we heard the "let's go," rumble, and some of the elephants did appear to begin moving off, towards the woods behind.

And then.

Suddenly, from the woods, we heard the most excruciating elephant trumpets and screams we've ever heard. They were yelling at the top of their voices!

Every single elephant in sight, including tiny babies, immediately turned tail and RAN into the woods. Within seconds they were gone. Simply disappeared. You couldn't even believe there had BEEN any elephants there a moment ago.

They had all left.

And so must we.

And we cried. Together.

"The cycle of elephant movements gave the bush a marked rhythm that bound us to its music. The elephants were like silent conductors of a natural orchestra, seeming to summon the frenzied frog calls of the wet season and the raucous hippo bellowing, all of which reverberated up and down the river at sunset. We soaked it in, drunk with a love for the land that ran so deep and strong that it scared us. We had discovered a life that we couldn't bear the thought of leaving."

Caitlin O'Connell, "The Elephant's Secret Sense"

Once again I am heartsick for Africa … the wild open spaces, the bush, the elephants, the sound of lions in the night …

Made in the USA
San Bernardino, CA
01 September 2016